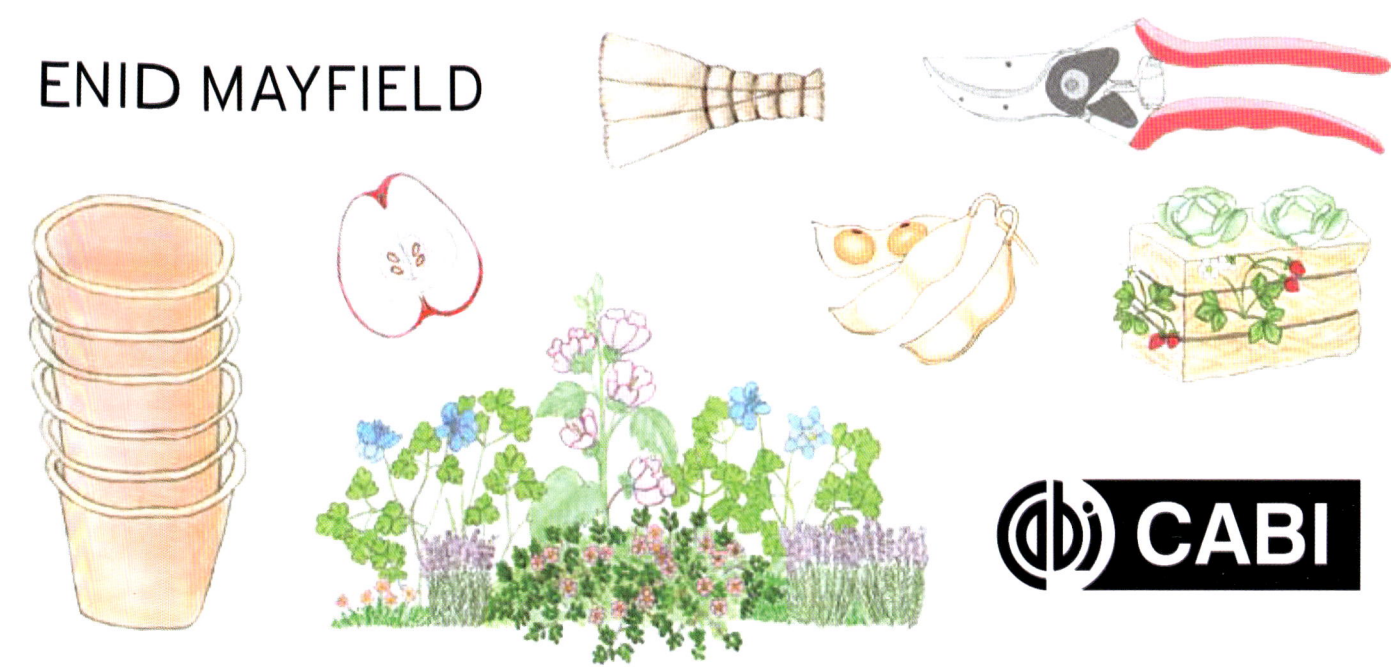

# ILLUSTRATED
# GARDEN GLOSSARY

ENID MAYFIELD

CABI

**CABI is a trading name of CAB International**

CABI
Nosworthy Way
Wallingford
Oxfordshire OX10 8DE
UK

Tel: +44 (0)1491 832111
E-mail: info@cabi.org
Website: www.cabi.org

CABI
200 Portland Street
Boston
MA 02114
USA

T: +1 (617)682-9015
E-mail: cabi-nao@cabi.org

A catalogue record for this book is available from the British Library, London, UK.

**Library of Congress Cataloging-in-Publication Data**

Names: Mayfield, Enid, author.
Title: Illustrated garden glossary / Enid Mayfield.
Description: Boston, MA : CAB International, [2025] | Includes
    bibliographical references. | Summary: "A comprehensive glossary of over
    1000 terms related to horticulture and gardening, with superb colour
    illustrations to aid comprehension"-- Provided by publisher.
Identifiers: LCCN 2024056046 (print) | LCCN 2024056047 (ebook) | ISBN
    9781800629691 (paperback) | ISBN 9781800629707 (pdf) | ISBN
    9781800629714 (epub)
Subjects: LCSH: Horticulture--Dictionaries. | Gardening--Dictionaries. |
    Dictionaries.
Classification: LCC SB45 .M39 20205  (print) | LCC SB45  (ebook) | DDC
    635.03--dc23/eng/20250219
LC record available at https://lccn.loc.gov/2024056046
LC ebook record available at https://lccn.loc.gov/2024056047

ISBN-13: 9781800629691 (paperback)
          9781800629707 (ePDF)
          9781800629714 (ePub)

DOI: 10.1079/9781800629714.0000

All illustrations are by the author
Cover design by Cath Pirret
Printed and bound in the UK by Bell & Bain Ltd, Glasgow

The paper this book is printed on is in accordance with the standards of the Forest Stewardship Council® and other controlled material. The FSC® promotes environmentally responsible, socially beneficial and economically viable management of the world's forests.

# Contents

# About the author

I was born in a small country town in South Australia. Our neighbour was a gentleman who had travelled to Europe in those days and as a child, I visited every day and chatted and watched while he pruned roses in the rosery and grafted fruit trees in the orchard. His housekeeper preserved fruit and made jams and marmalades that were stored away in a dark windowless pantry with a wooden floor.

My grandparents were farmers. There was an enclosed kitchen garden where Grandma grew lachenalias, snapdragons and poppies. The large front garden of the farmhouse was planted out with vegetables, the cow provided milk and cream, and the chooks laid eggs. Grandpa killed a sheep for meat, and at Christmas, Grandma chose the fattest chicken, chopped its head off, plucked it and removed the giblets. It was cooked in the wood oven and served for lunch on Christmas Day with bread stuffing, gravy, roast potatoes and vegetables.

When I was six we moved to Adelaide. I had a playhouse at the bottom of my parent's garden, with my own little garden. I grew radishes, broad beans and peas. A currant grape vine climbed over one of the wooden walls.

We built a house when I married. The garden had 20 fruit trees, including a quince, fig, stone fruits, apples and citrus, and I followed in the footsteps of my grandmother, the housekeeper and my mother and preserved fruit, and made jams and marmalade, and chutneys. We grew vegetable and strawberries in raised garden beds.

At our country property we had a large covered orchard with different kinds of plums, prunes, cherries and heritage apples. My square foot garden produced prolifically. Around the house we grew camellias and roses. There was always plenty to share and give to the local whole foods store.

We now live in subtropical Queensland and have a smaller garden. We grow papaya, avocado, grumichama and blueberries. The lemon and lime bear well, and we have two apple cultivars suited to the climate. And, of course, we have a herb garden.

There is no greater joy than preparing food sourced from one's own garden.

Our future in this country depends on it.

# Acknowledgements

I would like very much to thank the following people who generously gave their time and expertise to review different sections of this book:

Professor Tim Entwistle, former Director and Chief Executive of Royal Botanic Gardens Victoria
Dr. Sophie Parks, Senior Research Scientist, NSW Department of Primary Industries
Jonathan Lidbetter, East Coast Wildflowers, Mangrove Mountain, NSW
Jillian Becker, gardener and friend, South Australia

# Introduction

Following the publication of the *Illustrated Plant Glossary* in 2021, I began to think about other key terms that would be of use to gardeners. From this spark of an idea, I have created the *Illustrated Garden Glossary*; an accessible and appealing account of over 1000 terms related to gardening and horticulture.

This book represents both a comprehensive guide to key gardening terms, each supported by scientific full-colour illustrations, but also a journey into the history of our gardens across the worlds. Gardens can bring us such joy, whether it be through an afternoon in our own backyard, spent tending to flowers or vegetables, or to days wandering through the expansive beauty in botanic gardens, or the landscapes of stately homes.

This book covers many terms, including those relating to garden types, design, food production, permaculture, propagation, sustainable gardening approaches, soil, composting, planting and water use. I have also explored terms relating to the history of gardens, from ancient Greece and Egypt, through to gardens across Asia and the Middle East, and through to our modern-day urban farms.

I wanted this book to reflect my own love for gardening – and the pleasure that growing your own food in particular can bring. I also aimed to provide a glossary that is exceptionally thorough and useful for any individual who is passionate about gardening, or who works in or studies horticulture and plant sciences, farming or garden design, no matter their level of knowledge or where they are in the world. I hope that *Illustrated Garden Glossary* has succeeded in these aims.

# Themes

## COMPOST
aerobic composting
anaerobic composting
biodegradable
bokashi
brown compost material
chop and drop
cold composting
compost
compost accelerator
compost bin
compost booster
compost problems
compost starter
compost tumbler
composting worms
cover crops
dig and drop composting
green compost material
green manure
heap composting
hot composting
lasagne composting
leaf mould
mushroom compost
organic matter
pit composting
sheet composting
three-bin composting
trench composting
vermicomposting
worm composting
worm farming
worm wee

## GARDEN HISTORY
Al-Andalus
ancient Egyptian gardens
ancient Greek gardens
ancient Roman gardens
Baroque gardens
bonsai
botanic gardens
chahar bagh garden
chinampas
Chinese gardens
cloud pruning
cottage garden
food forest gardens
garden history
Hanging Gardens of Babylon
herb gardens

Islamic gardens
Japanese gardens
knot garden
medieval gardens
parterre
peristyle
Persian gardens
physic garden
qanat
simples
Three Sisters Garden

## PESTS AND DISEASES
aphids
biological pest control
black spot
blight
Bordeaux mixture
cabbage white butterfly
canker
chemical pest control
chlorosis
codling moth
damping off
diatomaceous earth
fruit fly
fungicide
gall
gummosis
herbicide
honeydew
insecticide
leafminers
mealy bugs
natural pest control
necrosis
neem oil
organic pest control
pesticides
pests and diseases of gardens
powdery mildew
pyrethrum
scale

## PLANT CLASSIFICATION
angiosperms
botanical name
common name
cultivar
dicotyledons
eudicots
F1 hybrid

family
genus
gymnosperms
hybrid, hybridisation
hybrid vigour
monocotyledons
plant kingdom
scientific name
species
subspecies
variety

## PLANT GROWTH
adventitious
alternate bearing
annual
biennial
biennial bearing
bine
blanching
blind
blossom drop
bolting
bramble
bush
bush vs climbing
bush vs vining
buttoning
cauliflory
chill factor
chlorophyll
climate zones
climbers
climbing roses
clump
coppicing
creepers
Critical Root Zone
crown
deciduous
determinate growth
drip line
epiphyte
evergreen
floricane
forcing
frost
frost hardy
frost tender
frost tolerant
girdling
habit

habitat
half-hardy
hardiness
hardy
herb
humidity
indeterminate growth
leader
life cycle
microclimate
perennial
photosynthesis
primocane
rambling roses
relative humidity
ring barking
sap
sapling
scramblers
shrub
sport
subshrub
sucker
tender
tendril
thatch
trailers
tree
twiners
variegated
vine
water sprout

## PLANT MORPHOLOGY

anther
axillary bud
bamboos
bark
berry
bract
bracteole
bud
cactus
calyx
cane
citrus
complete flower
corolla
cycads
dioecious
double flower
drupe
egg cell
ferns
filament
flower
frond
fruit

grasses
imperfect flower
incomplete flower
inflorescence
internode
lateral bud
leaf
legumes
male and female flowers
melons
monoecious
nectar
nectary
node
nodule
orchids
outer bark
ovary
ovule
palms
pedicel
peduncle
pepo
perfect flower
petal
petiole
pome fruit
rhizome
root tuber
roots
roses
runner
sepal
stamen
stem
stem tuber
stigma
stolon
stone fruit
style
succulents
terminal bud
tuberous roots

## PLANTS FROM SEEDS

acclimatisation
bottom heat
cloche
cold frame
cotyledon
dibber, dibble stick, dibbler
dormant
epigeal germination
etiolation
flats
germination
glass house
greenhouse

hardening off
heat mat
horticultural fleece
hot house
hotbed
hypogeal germination
legginess
medium
microgreens
over-seeding
plug
polytunnel
potting mix
potting up
pricking out
propagator
rockwool
scarification
seed leaf
seed raising mix
seed tray with cells
seedling
set
shade house
sowing seeds
sprouts
stratification
thinning out
transplanting
true leaves
vernalisation

## PLANTING SYSTEMS

aeroponics
aquatic gardens
balled and burlapped
banana circle guild
bare-rooted plants
biodynamic gardens
catch crops
clean cultivation
companion planting
crop rotation
cut and come again
double digging
ecological gardens
elevated garden beds
exotic gardens
fedge
guilds
hedge
hedgerow
heeling in
hugelkultur
hydroponics
indigenous gardens
indoor plants
intercropping

interplanting
introduced plants
keyhole gardens
labyrinth
lawn
market gardens
maze
monoculture
moon gardening
native gardens
naturalised plants
no dig gardens
organic gardens
ornamental gardens
permaculture
polyculture
potager
raised garden beds
square foot gardens
straw bale garden
succession planting
terrarium
urban farming
wicking beds

## PROPAGATION

air layering
approach grafting
asexual propagation
back bulb
bark grafting
basal plate
bisexual flowers
blind bulbs
bridge grafting
bud union
budding
budwood
bulb chipping
bulb scales
bulb scaling
bulb sectioning
bulbel, bulbil, bulblet
bulbs
callus
callus bridge
cambium
chip budding
chitting
cleft grafting
clone
compound layering
corm
cormel
cross-pollination
cuttings
division
double working

eye
fertilisation
genetically modified seeds
graft
graft chimaera
graft compatibility
graft hybrid
graft union
graftage
grafting
green-grafting
greenwood cuttings
hand pollination
hardwood cuttings
heartwood
heel cutting
heirloom
hen and chicken fern
herbaceous cuttings
imbricate bulbs
inarching
inner bark
interstock
keiki
layering
leaf cuttings
leaf-bud cutting
live stake cuttings
mallet cutting
marcotting
micropropagation
mound layering
mutation
offset
open pollination
patch budding
phloem
piggyback plant
pole cuttings
pollen
pollination
pollinator
propagation
propagule
pseudobulb
pup
repair grafting
root cuttings
rooting hormone
rootstock
saddle grafting
sapwood
scaly bulb
scion
scooping
scoring
seed
seed potato

seed saving
seed tuber
selective breeding
self-fruitful
self-pollination
self-unfruitful
semi-hardwood cutting
semi-ripe cuttings
separation
serpentine layering
set
sett
sexual propagation
side grafting
simple layering
slip
slipping
softwood cuttings
splice grafting
split vein propagation
spore
stem cuttings
stock
stool layering
T-budding
tiller
tip layering
tissue culture
top grafting
topworking
trench layering
true to type
tuber
tunicate bulbs
twin-scaling
unisexual flowers
vegetative propagation
wedge grafting
whip and tongue grafting
wood
wounding
xylem

## PRUNING

cordon fruit trees
cutting back
deadheading
disbudding
espalier
festooning
heading back
hedging
pinching
pleaching
pollarding
pot bound
pruning
rejuvenation pruning

renewal pruning
root pruning
root-bound
shearing
spur
standard plant
thinning
tip pruning
topiary
topping
undercut pruning

## SOIL

acidic soil
aeration
alkaline soil
amendments
anaerobic soil
arthropods
bacteria
bedrock
biochar
blood and bone
bone meal
boron
calcium
chelation
chloride
clay
clayey soil
cobalt
coco peat
coir
complete fertiliser
copper
cover crops
crumb
earthworms
erosion
fertiliser
foliar feeding
friable soil

fungus
granular fertiliser
green manure
greensand
gypsum
hardpan
humus
hydrophilic
hydrophobic
iron
leaching
lime
litmus paper
loam
macronutrients
magnesium
manganese
manure
micronutrients
molybdenum
mulch
mycorrhiza
nematodes
neutral soil
nickel
nitrogen (N), ($N_2$)
nitrogen fixation
NPK ratio
organic fertiliser
organic matter in soil
parent material
peat moss
perlite
pH
phosphorus
plant nutrients
potash
potassium
protists
salinity
sand
sandy soil

seaweed
silicon
silt
silty soil
slow release fertilisers
soil
soil aggregates
soil biota
soil carbon
soil food web
soil pH
soil profile
soil structure
soil texture
soil wettability
solarisation
sour soil
sphagnum moss
subsoil
sulfur
sweet soil
tillage
tilth
topdressing
topsoil
trace elements
vermiculite
zinc

## WATER

black water
dams
grey water
groundwater
irrigation
micro-irrigation
misting
qanat
swale
water
water harvesting
water tank

**acclimatisation** *see* hardening off

**acidic soil**

Acidic soils are often referred to as 'sour'. They have a pH of less than 7.

Strongly acidic soil with a pH of less than 5.5 results in poor plant growth. It decreases the availability of nutrients to plants. Aluminium and manganese levels rise to a level that is toxic to plants. There are many causes of soil acidification, including the excessive application of ammonium-based fertilisers, deforestation, and land use practices that remove all harvested material.

Many plants like azaleas and rhododendrons (*Rhododendron*), prefer an acidic soil between 4.5 and 6.0. However, a neutral soil pH between 6.0 and 7.5 is generally acceptable for most plants.

*cf.* **alkaline soil**

**adventitious**

Of plants parts, like flowers and roots, growing in unusual places.

Examples include redbud (*Cercis*), and its cultivars, that grow flowers out of their woody trunks and branches. Ivy (*Hedera helix*) grows roots along its stems that help it cling to and climb along a surface.

Flowers growing from the branch of a redbud

Roots growing from the stem of ivy

**aeration**

Aeration is the process of puncturing compacted soils so that water, nutrients and oxygen can enter.

This can be done by making holes in the upper layer with a spike aerator or by removing cylindrical plugs of soil.

Soil compaction requiring aeration is a familiar problem with lawns.

**aerobic composting**

Aerobic composting of biodegradable material occurs in an environment with oxygen.

Microorganisms that thrive in an aerobic environment break down organic matter, like plant material. The time it takes depends on the method of composting used.

*see also* **compost**
*cf.* **anaerobic composting**

**aeroponics**

Aeroponics is a method of growing plants without soil and with little water. In aeroponics, roots are suspended in the air in a nutrient-rich mist that supplies all the growing plant's needs. Plants also need natural or artificial light and stable air temperature and humidity.

Commonly grown plants include leafy greens, culinary herbs and strawberries. Root vegetables, like carrots, are also suitable. Systems can be built to allow for vertical growth of climbers like beans, cucumbers and tomatoes.

AEROPONICS

LED light

Seeds are planted in
a special medium
that supports the
plant as it grows

Plant support with
holes for pots

Roots are suspended
in the air and misted
with a nutrient-rich
solution

Water
circulation
pump

Nutrient
solution

## air layering

Air layering, also called marcotting, is a form of propagation.

A branch or branchlet is girdled and enclosed in a moist rooting medium. Moisture and nutrients that gather at the girdling site encourage root growth. Once roots have formed, the branchlet can be cut away from the mother plant and potted up.

PROPAGATION BY AIR LAYERING

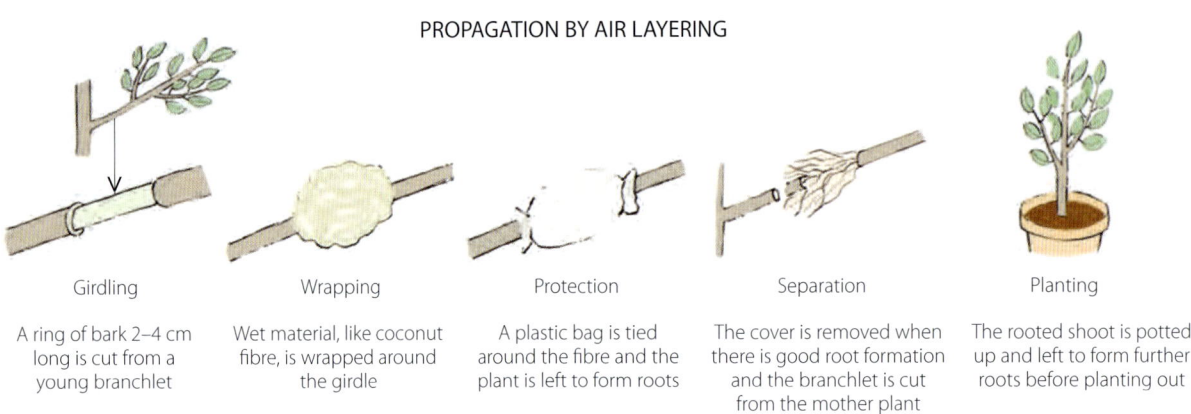

| Girdling | Wrapping | Protection | Separation | Planting |
|---|---|---|---|---|
| A ring of bark 2–4 cm long is cut from a young branchlet | Wet material, like coconut fibre, is wrapped around the girdle | A plastic bag is tied around the fibre and the plant is left to form roots | The cover is removed when there is good root formation and the branchlet is cut from the mother plant | The rooted shoot is potted up and left to form further roots before planting out |

## Al-Andalus

Al-Andalus is the Muslim kingdom that occupied Spain and Portugal from 711 to 1031 AD. The kingdom then fragmented from the 13th century and was taken over by Christians in the 15th century.

Theirs was an advanced and learned culture. Their intensive irrigation system transformed previously uncultivated land or land with poor yields. They understood grafting techniques, even grafting plants of different genera like the apple (*Malus domestica*) and the almond (*Prunus dulcis*). They researched soils, fertilisers, and acclimatising new plants like the banana.

Beauty in gardens reflected God and fed the spiritual life. Even domestic gardens had plants like jasmine. The wealthy had a refined, elegantly designed ornamental garden with running water and fountains, a remnant of which is in the Alhambra Palace in Granada.

Peasants worked market gardens, fruit orchards, vineyards and olive groves, as well as their own plots of garden. Orchard gardens had aromatic plants and vegetables, as well as fruit trees.

*see also* **Islamic gardens**

The irrigation system was based on channels of different sizes. Rivers were dammed, and water was distributed according to need on a strict schedule.

*After detail from a contemporary image of a garden in Al-Andalus*

### alkaline soil

Alkaline soils are often referred to as being 'sweet' or limey. They have a pH of more than 7.

High alkalinity in soils (with a pH of greater than 8.5) is devastating for plants because it causes nutrients to become unavailable. The soil structure collapses and loses the porosity necessary for the dispersal of water and gases.

Some plants, like beetroot, leeks and spinach, prefer an alkaline soil between 7.0 and 8.0.

Generally, a neutral soil pH between 6.0 and 7.5 is acceptable for most plants.

*cf.* **acidic soil**

### alternate bearing

Alternate bearing, also called biennial bearing, refers to the tendency of a fruit tree to bear fruit in a 2-year cycle. A large crop one year is followed by a small crop the next year.

While fruit growth for this year's crop is well underway, next season's flowers are forming on shoots with next year's buds. This occurs before the tree becomes dormant.

If the tree's energy is used for fruit development, little energy remains for flower bud development. This results in fewer flowers and less fruit the following season.

### amendments

Amendments are materials added to soil to improve its texture, adjust pH, add nutrients, and/or improve its water-holding capacity. They include compost and worm compost, well-rotted animal manure, cover crops and greensand. Other amendments are seaweed and seaweed extracts, blood and bone, and lime.

### anaerobic composting

Anaerobic composting of biodegradable material occurs in an environment that is starved of oxygen.

Microorganisms that thrive in an anaerobic environment break down organic matter, like plant material and kitchen scraps, in an airtight environment over a relatively long time.

Trench, pit and bokashi composting are examples of anaerobic composting.

*see also* **compost**
*cf.* **aerobic composting**

## anaerobic soils

Anaerobic soils are waterlogged or poorly drained. Open spaces in the soil are replaced with water and free oxygen is deficient. Root growth and absorption of nutrients is restricted.

Oxygen is vital for soil organisms and cannot enter anaerobic soils. Oxygen, carbon dioxide and water vapour all diffuse into and out of the soil from or to the atmosphere through open spaces in the soil.

Good organisms die and are replaced by toxic microbes. Plants in anaerobic soil have slow leaf and shoot growth, yellowing of older leaves, wilting, and disease.

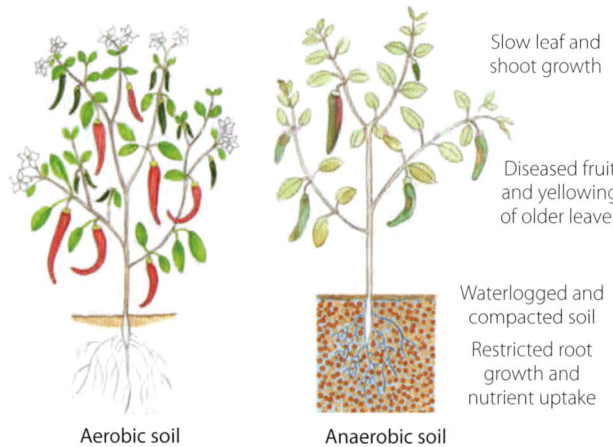

Slow leaf and shoot growth

Diseased fruit and yellowing of older leaves

Waterlogged and compacted soil

Restricted root growth and nutrient uptake

Aerobic soil        Anaerobic soil

## ancient Egyptian gardens *see* page 5

## ancient Greek gardens

Ancient Greece spanned the period from the 8th century BC to the Roman conquest in 146 BC.

We know about Greek gardens from the fictional works of Homer (*The Iliad* and *The Odyssey* in the 8th century BC), and the botanical and horticultural treatises of Theophrastus (372–287 BC) who is considered to be the father of botany.

In urban areas houses were crowded together around a public square planted with shade trees. There was no space for a private garden in the courtyard. Instead, plants in pots were placed on the rooftop. Homer mentions an enclosed sacred grove garden, with an altar and shade-spreading trees, where people went to worship the gods.

Apart from these public spaces, gardens were utilitarian. Outside the city walls farmers grew grain, olives, grapes, fruit trees (including figs, apples and pears) and vegetables (including onions, beans, garlic and lentils). Flowers were grown for weddings and festivals. Roses were known from the earliest times and were cultivated extensively. Peaches and lemons were introduced from Asia in the time of Theophrastus.

**ANCIENT GREEK GARDENS**

Harvesting olives
Detail from an attic amphora c. 520 BC.

Garden pots to be placed on the rooftop
Detail from an oil jar c. 425–375 BC

## ancient Egyptian gardens

The highly sophisticated kingdoms of ancient Egypt, from 3000 BC to 100 AD, had cultivated gardens.

Temple gardens, palace gardens, estate gardens and domestic gardens were all within a walled compound. They were watered from the River Nile by a system of irrigation canals. The vast majority of people were involved in farming that took place outside the compounds along the banks of the River. After the annual floodwaters receded, crops were planted in the fertile alluvial soil deposited by the River and harvested before the beginning of the next flood cycle.

Nebamun was an accountant at the temple of Karnak who died in c. 1350 BC. An image of his estate garden decorated the chapel of his tomb. The garden is both elegant and useful. It shows the central rectangular water feature surrounded by shade trees that also produced food. The garden is symmetrical, with trees and plantings mirrored on each side around the pool. Each has a symbolic sacred meaning.

ANCIENT EGYPTIAN GARDENS

doum palm
(*Hyphaene thebaica*)

The fruit is edible and was placed in tombs for sustenance in the afterlife.

sycamore fig
(*Ficus sycomorus*)

A valued shade tree. The fig grows on the trunk and branches and is edible.

date palm
(*Phoenix datilyfera*)

It was associated with the sun god Ra because its tall stem and leaves reached to heaven. Its sweet fruit was prized.

blue lotus
(*Nymphaea caerulea*)

A symbol of rebirth and fertility. Roots, seeds, fruits and petals were all useful.

mandrake
(*Mandragora officinarum*)

A hallucinogenic plant containing atropine and thought to be included in sacred rituals.

papyrus
(*Cyperus papyrus*)

An aquatic sedge used for making paper and boats. The pith and rhizomes are edible.

corn poppy
(*Papaver rhoeas*)

Grew as a weed among crops. Cultivated flowers were used in bouquets for the gods.

cornflower
(*Centaurea depressa*)

Grew as a weed among crops and was cultivated in gardens. A symbol of life and fertility.

## ancient Roman gardens

The ancient Roman state began in the 8th century BC in the city of Rome, and lasted until the 5th century AD. During that time, it occupied much of Europe, North Africa and the Middle East. It spread its own culture and integrated others.

There were serene sprawling temple gardens, and large pleasure gardens with ornate terraces for public use.

Private gardens were symmetrical enclosed, as were those of the Egyptians, Greeks and Persians before them.

In wealthy Roman homes there were more lavish peristyle gardens with a covered walkway, supported by colonnades, surrounding the open courtyard garden. It had formal beds, statues, water features, flowers, like roses and narcissus, trees and topiaried shrubs. In the Middle Ages this design evolved to become the more mundane monastic cloister.

The Romans had fruit, herb and vegetable gardens. These practical gardens remained a feature of the country villas of the wealthy, and by the 1st century they had hothouses for forcing fruits. Orchards were planted using the quincunx or diagonal planting system that pleased the Roman sense of order, as from every direction you looked at the orchard you saw many straight lines of trees.

### ANCIENT ROMAN GARDENS

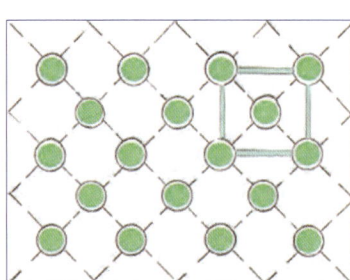

A quincunx planting
is based on a square with four plants at
each corner and an additional plant in
the centre.
This creates a design with many
diagonal rows.

The open inner courtyard of a house with a peristyle garden.
The surrounding covered walkway is supported by colonnades.
Garden frescoes on the walls extend the sense of space.

---

**angiosperms** *see* page 7

**annual** *see* life cycle

**anther** *see* flower

**aphids** *see* biological pest control, honeydew, pests and diseases of gardens, pyrethrum, white oil

# angiosperms

Angiosperms are seed-bearing plants that have flowers. They include herbaceous plants, shrubs, grasses, and trees. There are two groups of seed-bearing plants, the other being gymnosperms, that have cones instead of flowers.

The reproductive structure of an angiosperm is in a flower, and comprises an ovary with ovules, and anthers with pollen. Seeds develop enclosed in an ovary that becomes a fruit.

Currently, angiosperms are classified into several groups, the most common of which are eudicots (75%), and monocotyledons (22%). Eudicot seeds usually have two seed leaves (cotyledons) and monocotyledons have one.

ANGIOSPERMS

MONOCOTYLEDONS

EUDICOTS

bulbs grasses palms

herbs trees shrubs

Usually herbaceous plants that lack side shoots

Herbaceous or woody plants

SEEDS

**Seed** with one seed leaf (cotyledon)

cotyledon

onion

cotyledons

**Seed** with two seed leaves (cotyledons)

rock cress

FLOWER PARTS

crocus lily

**Flower parts** usually in multiples of 3. Sepals often same colour and shape as petals.

geranium pea

**Flower parts** usually in multiples of 4 or 5. Sepals and petals are usually distinct.

LEAF VENATION

Leaves with **parallel main veins**

Leaves with **lattice-like veins**

ROOTS

**Roots** usually fibrous, never a taproot

**Root system** often a taproot

## approach grafting

During approach grafting, both the rootstock and the scion remain attached to their own root systems.

A slice of bark is removed from both the scion and the rootstock and the two exposed areas are bound together securely to ensure close contact.

After the graft union forms, the rootstock is severed from the scion, and the part of the rootstock above the graft is cut off.

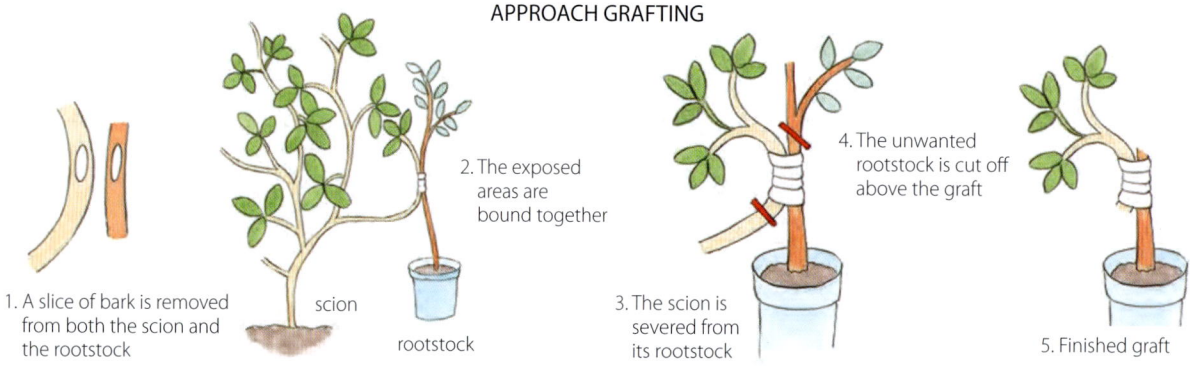

APPROACH GRAFTING

1. A slice of bark is removed from both the scion and the rootstock

scion

rootstock

2. The exposed areas are bound together

3. The scion is severed from its rootstock

4. The unwanted rootstock is cut off above the graft

5. Finished graft

---

## aquatic gardens *see* page 9

## arthropods

Arthropods include spiders, centipedes, mites, millipedes, butterflies, moths and insects. They typically have a segmented body and a hard, outer skeleton that is shed from time to time to enable growth.

Arthropods in the soil break down dead plant and animal residues. Fungi and bacteria feed on their nutrient-rich excreta and, in the process, make it available to plants. Mites benefit the soil and prey on soil-borne pests.

Arthropods are either beneficial predators like spiders, or pests, like some wasps.

mite                     millipede

Soil mites break down dead plant and animal residue. Millipedes also break down dead plant material but can become a pest when they turn to eating leaves, stems and roots.

## asexual propagation = vegetative propagation

## axillary bud *see* bud

## back bulb

A back bulb is the old, leafless pseudobulb of some orchids that remains after the above-ground growth is finished.

If the back bulb is alive and has a bud at the base, it can produce a new pseudobulb.

pseudobulb

back bulb

new pseudobulb

back bulb

If the back bulb is alive, it is removed and can produce a new pseudobulb

## aquatic gardens

Aquatic plants live and grow in wet environments. They are found in all types of water, whether freshwater or seawater. They are placed in categories based mainly on where they like their roots and leaves to grow.

In gardening, they are used around and in ponds, water features and dams. They may have an ornamental use, as well as adding oxygen to the water. Aquatic plants keep the water clear and clean. They provide habitat for fish and frogs. For the best growth, plants need to be placed at an appropriate depth.

### AQUATIC PLANT GROWTH

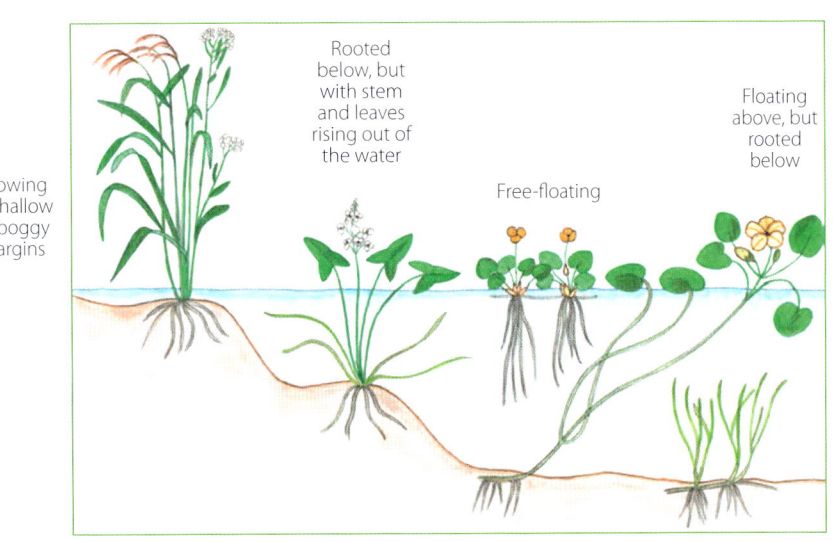

| | |
|---|---|
| **Shallow or boggy margins** | Plants that grow where the land and water meet. Depth may be 0–15 cm. |
| | creeping Jenny (*Lysimachia nummularia*), aquatic ginger (*Alpinia aquatica*) |
| **Deep margins** | Plants that usually like 15–30 cm of water, bottom rooted with floating leaves and flowers. |
| | Pickerel weed (*Pontederia cordata*) |
| **Deep-water plants** | Grow in water from 30 cm–1.2 m or more deep, bottom rooted with floating leaves and flowers. |
| | arrowheads (*Sagittaria sagittifolia*), yellow floating heart  (*Nymphoides peltata*) |
| **Submerged plants** | Grow entirely under water. These are the best oxygenators. |
| | ribbon grass (*Vallisneria spiralis*) |
| **Free-floating plants** | Plants that float on the pond's surface and do not root in the soil. |
| | golden bladderwort (*Utricularia aurea*) |

### SOME NATIVE AQUATIC PLANTS

swamp goodenia
(*Goodenia humilis*)

river mint
(*Mentha australis*)

matted pratia
(*Lobelia pedunculata*)

water plantain
(*Alisma plantago-aquatica*)

swamp iris
(*Patersonia fragilis*)

## bacteria

While bacteria may be small, about the size of a clay particle (< 0.002 mm), they make up both the largest number and weight of any soil microorganism.

Bacteria decompose organic matter into forms useful to themselves and other organisms in the soil food web.

Some partner with plants in a way that is mutually beneficial. Nitrogen-fixing bacteria in legumes and some other plants, like alder, receive nutrients from the plant and in turn fix nitrogen from the air into a form the plant can use.

Other bacteria cause plant diseases, like galls.

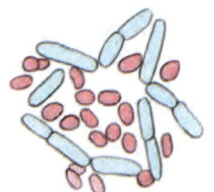

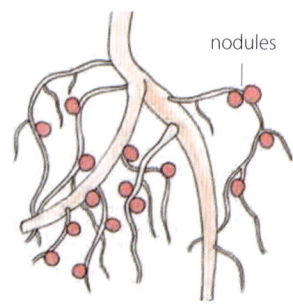

Bacteria make up the largest number and weight of any soil microorganism

*Rhizobium* bacteria in the root nodules of legumes fix nitrogen

## balled and burlapped

Various types of trees and shrubs are grown in a nursery field for a limited time. When they are lifted, the roots and the soil surrounding them are wrapped in burlap or another coarse fabric. The covering is tied and secured with twine or wire.

The wrapped plant is now ready for sale and transplanting to the garden.

Plants sold this way include some deciduous trees and shrubs, evergreen trees like citrus and conifers, and shrubs like azaleas and rhododendrons

## bamboos

A bamboo is a usually tall, tree-like grass, with hollow woody stems (culms).

There are two types of bamboo. Most gardeners choose a clumping bamboo rather than a running bamboo that can easily become invasive.

China gold (*Bambusa eutuldoides* var. *viridi–vittata*) is a clumping bamboo used as a screen. The dwarf green stripe (*Pleioblastus viridistriatus*) is a running bamboo with a maximum height of 50 cm that works well as a ground cover. Some bamboos, including running bamboos, can be grown in pots.

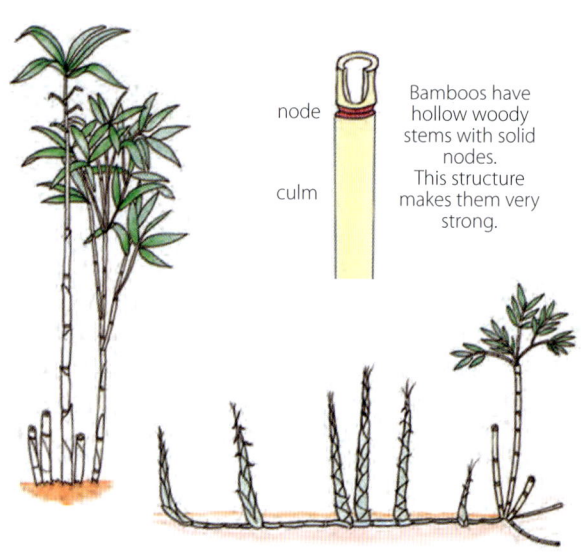

Bamboos have hollow woody stems with solid nodes. This structure makes them very strong.

node

culm

## banana circle guild

A banana circle guild is a circular garden about 2 m wide with a hole about 1 m wide dug in the middle.

The soil from the hole is mounded around the perimeter to about 60 cm in width. Bananas are planted on the mound with other food species like the taller shade-giving paw paw, ground covers like sweet potato and perennial peanut (also a nitrogen fixer), cassava and taro.

The central hole is filled with organic matter that rots down and creates compost to feed the plants.

Banana circle guilds are suited to the tropics and subtropics.

### BANANA CIRCLE GUILD

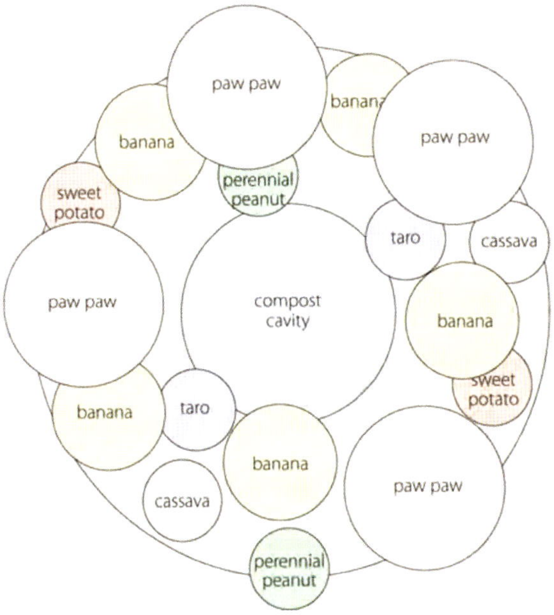

paw paw
banana
banana
perennial peanut
sweet potato
paw paw
taro
cassava
paw paw
compost cavity
banana
sweet potato
banana
taro
banana
paw paw
cassava
perennial peanut

compost

1 m
2 m

paw paw
(*Carica papaya*)
6–8 m high.
Fruit edible.

taro
(*Colocasia esculenta*)
to 1.5 m high.
Corms edible.

banana
(*Musa*)
5 m high.
Fruit edible.

perennial peanut
(*Arachis glabrata*)
15 cm high.
Nitrogen-fixing legume.

cassava
(*Manihot esculenta*)
1–5 m high.
Edible tubers grow from roots at the base of the stem

sweet potato
(*Ipomoea batatas*)
Vine spreading to 6 m.
Roots with edible tubers.

## bare-rooted plants

During winter, dormant deciduous trees and shrubs that are grown in a nursery field are dug up.

Little or no soil is left on the roots, and they are sold bare-rooted rather than potted up.

Nearly all deciduous plants can be bare-rooted for sale, including fruit trees and roses.

Bare roots must be kept moist before planting. The tree or shrub is watered in when planting and again only when dormancy ends.

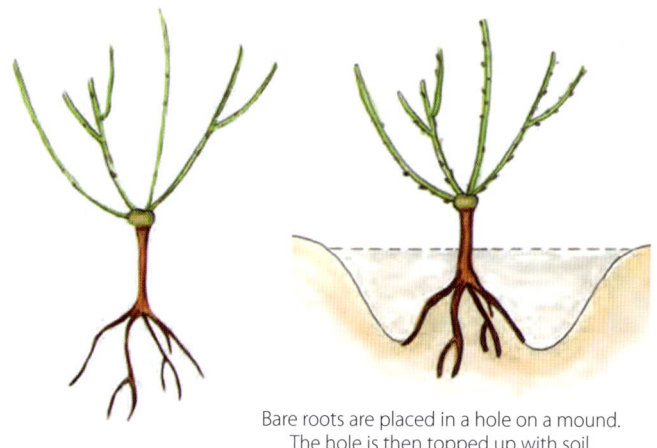

Bare roots are placed in a hole on a mound. The hole is then topped up with soil.

## bark

Bark is the outer layer of a woody stem or root. It consists of the dead outer bark and the living inner bark (phloem).

## bark grafting

Bark grafting is used when the rootstock of an established tree is too big for the scion.

Several scions are distributed under the bark around the circumference of the rootstock. The cambium of the scion and rootstock must be in contact. The graft is then sealed to prevent the scions and limb from drying out. As they get bigger, some of the new shoots are removed.

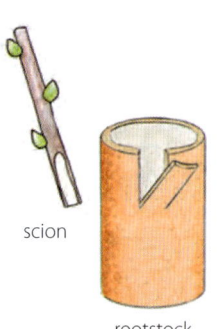

scion

rootstock

Scions are inserted into the rootstock and nailed in place

The graft is then protected with a layer of wax

## Baroque gardens

This formal garden style evolved in Italy and France in the early 17th century. It dominated European garden design until the mid-18th century. These were immense gardens dominated by parterres, long walkways, and borders of trees. They were a mark of the status of the owner.

The most famous Baroque gardens are the gardens at the Palace of Versailles in France. They were designed by André Le Nôtre between 1662 and 1666.

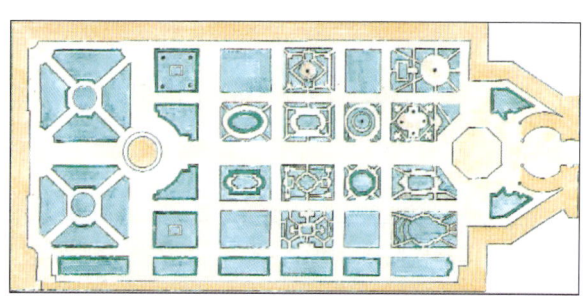

C17th design in the Baroque style, by André Le Nôtre, for the Tuileries Gardens in Paris

## basal plate

A basal plate is a compressed stem at the base of a bulb or corm from which fleshy scales, leaves, flowering stems, roots and offsets grow.

Both tunicate bulbs and scaly bulbs have a basal plate.

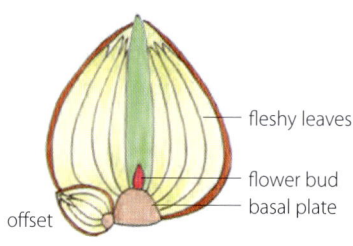

fleshy leaves

flower bud

basal plate

offset

onion bulb (*Allium cepa*)

## bedrock

Bedrock is a solid mass of rock below the soil layers.

## berm

A berm is a mounded, sinuous wall of soil that is usually about 0.5 m high. It is built on a gradual slope and planted out so that it remains stable.

Larger berms are created as features in a flat landscape. A small berm can be constructed around the base of a tree to hold water when watering.

*see also* **swale**

## berry

A berry is a small, pulpy fruit that is often edible.

Grapes, cranberries and blueberries are simple berries. Many fruits not commonly known to be berries include bananas, aubergines, tomatoes and kiwifruit.

Raspberries and blackberries are clusters of tiny drupes (drupelets) and are not berries at all.

Grapes, aubergines and tomatoes are berries

**biennial** *see* life cycle

**biennial bearing** *see* alternate bearing

## bine

A bine is the flexible twining stem of some plants, such as hops (*Humulus lupulus*), bindweed (*Convolvulus*) and runner beans (*Phaseolus coccineus*).

It climbs by spiralling its stem around another plant for support as it grows upward. Stiff hairs on the stem help anchor it.

Bine is also the name of a plant with such a stem.

hops

bindweed

## biochar

Biochar is a soil additive that improves soil health and fertility.

It is a charcoal-like material produced when plant matter, manure or other organic material is burned slowly in a low-oxygen environment. This prevents the carbon in the material from combining with oxygen to form carbon dioxide, which would then be released into the air. The process captures carbon in a solid form that is called biochar.

Biochar must be activated before it is placed in the garden. The charred remains are crushed, then activated by mixing them with water and worm castings, garden compost, organic fertilisers or liquid kelp. After a few days, it can be added to the garden. Adding biochar to soil is a way of storing carbon in the ground that would otherwise be released into the atmosphere.

Good quality biochar improves soil structure and, because it is highly porous, helps retain soil moisture. Root bacteria and mycorrhizal fungi grow in the pores, and micronutrients that can be made available to plants bond loosely to the surface of the pores.

There are many ways to make biochar. It is also available commercially.

## biodegradable

Biodegradable describes substances that can be decomposed naturally by bacteria, fungi or other living organisms.

## biodynamic gardens

Biodynamic gardens are organic gardens with additional features like the observation of planetary influences, such as moon phases and zodiac signs, to determine the best time to do garden tasks.

These gardens aim to be entirely self-supporting. An ideal biodynamic garden will be a closed loop system that brings nothing in from outside. Neither an organic gardens nor a permaculture garden can be a closed loop system because garden supplies like hay, manure and seeds are brought in from outside the property.

## biological pest control

All organisms have natural enemies like predators that kill and eat them, parasites that lay their eggs inside them, and viruses, bacteria and fungi that cause infectious diseases. Biological controls use natural enemies to suppress, but not eliminate, plant pests. They target the pest and cause no damage to the plants themselves, and leave no residues.

Some biological controls are available to home gardeners. Ladybirds help control leafhoppers, mealy bug and aphids. *Bacillus thuringiensis* is a naturally occurring bacterial disease of insects that is often used in pesticides. It controls leaf-feeding caterpillars, like those of the cabbage white butterfly that lays its eggs on brassicas.

Other control pests can be purchased by home gardeners. Cucumeris (*Neoseiulus cucumeris*) are predatory mites that are effective against some thrips. Persimilis (*Phytoseiulus persimilis*), another predatory mite, feeds on two-spotted spider mites and other spider mite species.

## bisexual flowers

A bisexual flower has both male (stamens) and female (pistils) reproductive organs in the same flower.

A bisexual flower may be able to both self-pollinate and cross-pollinate. Many bisexual flowers have mechanisms to prevent self-pollination. For example, pollen may mature before the stigma is ready to receive it.

One pistil and four stamens
mint (*Mentha*) flower

Many pistils and stamens
buttercup (*Ranunculus*) flower

*cf.* unisexual flowers

**black spot** *see* Bordeaux mixture, pests and diseases of gardens

## blackwater

Blackwater is waste water from toilets and bathrooms that contains urine or faecal matter. Kitchen and dishwasher water are also considered blackwater due to contamination by pathogens and grease.

Black water is also known as sewage and can transmit diseases and bacteria that can be harmful to health. It must pass through a treatment system. Treated blackwater is sometimes used for irrigation.

*see also* grey water, groundwater

## blanching

Blanching is used to block sunlight from plant parts to prevent them from greening.

Leaves are used to cover the head of a cauliflower so that it stays creamy-white.

Soil is heaped against the stems of leeks and celery as they grow to keep them white. It can also improve flavour by preventing bitterness.

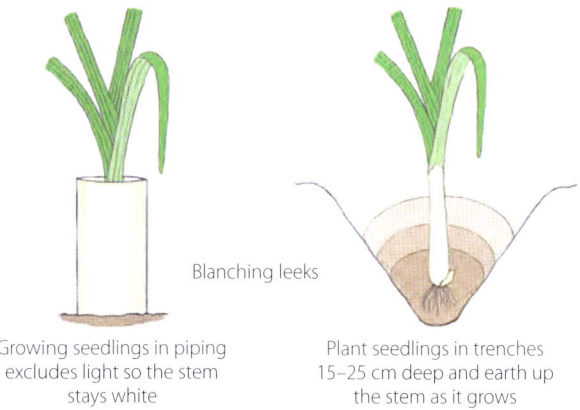

Blanching leeks

Growing seedlings in piping excludes light so the stem stays white

Plant seedlings in trenches 15–25 cm deep and earth up the stem as it grows

## blight *see* pests and diseases of gardens

## blind

Plants, like some daffodils, that should bloom but don't are said to be blind.

## blind bulbs

Blind bulbs, like those of some daffodils, have foliage but produce no flowers.

## blood and bone

Blood and bone is a natural fertiliser that is a by-product of the slaughter of cattle. It had been used for centuries.

Blood meal, that is ground, dried and boiled blood, provides nitrogen, and ground bone meal provides mainly phosphorus and some calcium. Fish meal provides phosphorus, nitrogen and potassium.

## blossom drop

Blossom drop refers to flowers that fall without bearing fruit. It is often due to high temperatures, stress, or lack of pollination. It is often due to pests like thrips, high temperatures, too much or too little water, or lack of pollination. Male flowers drop naturally from vegetable plants after a few days.

## bokashi

Bokashi is an anaerobic process that ferments kitchen waste, including meat and dairy, into a soil builder. The by-product is a nutrient-rich tea. The tea is quite acidic and should be diluted before being using it to feed plants.

The waste is mixed with a bran inoculated with microbes that flourish in an oxygen-starved environment. It is then placed in an airtight bin with a spigot for draining the liquid tea. Chop the waste items before placing them in the bin for more effective fermentation. Once the bin is full, allow it to continue fermenting for about a month. The waste will not be broken down, but the pickled product can now be buried in the soil or added to a traditional compost pile.

It is possible to use two bins, one for current waste and another for waste that is continuing to ferment.

Bokashi was developed in the early 1980s by Dr. Teuro Higa, at the University of the Ryukyus in Okinawa, Japan.

BOKASHI

Inoculated bran is sprinkled over the layers of food waste as they are added

airtight lid

spigot

cup for collecting liquid

Bin insert with holes separates solids from liquids

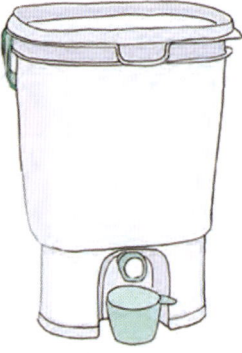

After a few weeks, the material in the second bin is a pickled 'pre-compost' that will break down more quickly in the soil or in compost

## bolting

Bolting, also called going to seed, is the premature flowering of a plant, usually a vegetable, that sets seed before the crop can be harvested.

Its causes can be longer and warmer days that trigger flowering, late planting of crops like brassicas that need cool temperatures for growth, a lack of water, or other stresses. At these times, the plant's priority is to quickly set seed so it can reproduce.

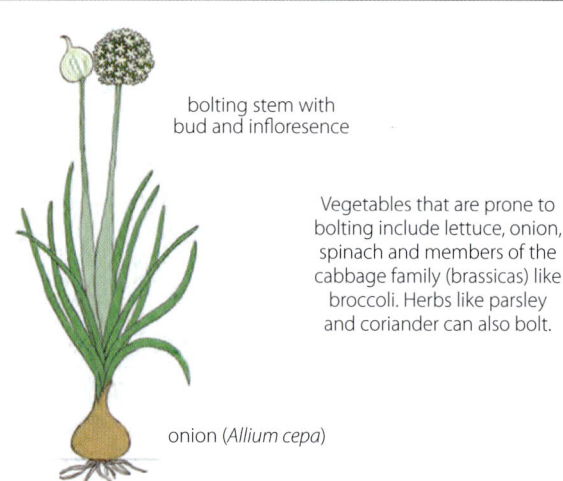

bolting stem with bud and infloresence

Vegetables that are prone to bolting include lettuce, onion, spinach and members of the cabbage family (brassicas) like broccoli. Herbs like parsley and coriander can also bolt.

onion (*Allium cepa*)

**bone meal** *see* blood and bone

**bonsai** *see* page 17

## Bordeaux mixture

Bordeaux mixture is a fungicide that was developed in the Bordeaux region of France for use before downy mildew appears on grape leaves and fruits.

It is a combination of copper sulphate, lime and water, and is an approved organic fungicide, bactericide and algaecide.

Fungal diseases treated with Bordeaux mixture include apple scab, peach leaf curl, canker, and black spot on roses. Instructions for use should be followed carefully.

**boron** *see* plant nutrients

## bonsai

The ideal bonsai is a single tree. It looks like a normal tree, but in a miniature form. It is shaped and designed in a way that its age is exaggerated with no visible signs of any work done on the tree. Trees need to be a certain height and to be balanced, though not symmetrical. Years of constant pruning and attention are devoted to them. Wire is used to bend a tree to a desired shape and roots are trimmed to limit size.

There are a number of set designs. An upright style reflects the basic concept of a tree with the trunk pointing upwards, thick branches at the bottom and thinner branches at the top. Another style is similar but incorporates more natural-looking curves in the trunk. Slanted styles have the tree leaning in one direction, and windswept styles make the tree appear as if it has been shaped by a strong wind.

Bonsai first appeared in Japan in the 12th century. It reflects the tenets of Zen Buddhism. *Wabi-sabi* finds beauty in transience and ageing over time, and *shibusa* is associated with unobtrusive beauty. Special techniques are used to achieve this. *Jin*, a bare-stripped part of a branch, creates deadwood that reflects the stresses of nature. Pots are simple.

Bonsai, refers to a pruned potted miniature tree, as distinct from *niwaki*, or cloud pruning, that refers to a garden tree.

Bonsai has its own set of specifically designed traditional tools.

BONSAI

Deadwood (jin) is prized

Balance is sought rather than symmetry

*Bonsai* means 'tray planting'

Bonsai reflects 'unobtrusive beauty'.
oak (*Quercus*)

coir fibre moss brush

spherical knob cutter

pruning shears

wire cutters

jin and wire pliers

Each task has the appropriate tool

## botanic gardens

Botanic gardens are a scientifically based collection of living plants that are used for botanical and horticultural research. They are places for plants to be conserved and are popular for recreation.

In Europe, medicinal herb gardens (physic gardens) were connected with the medical faculties of the earliest universities. These became the first botanic gardens. Both Florence and Padua established gardens in 1545.

The two centuries from 1700 to 1900 saw the expansion of the British, French, Spanish, Portuguese and Dutch empires into the New World and the collection of vast quantities of plant material that were taken back to European botanic gardens to be studied.

Botanic gardens were also established in the colonies. They acted as collecting stations for plants destined for Europe. They were also places for trialling plants to find those best suited to the local climate. Among them were the botanic gardens established by the Portuguese in Rio de Janeiro, Brazil (1808), and by the British in Singapore (1859), and in Sydney, Australia (1816).

### GEELONG BOTANIC GARDENS

The 21st century garden has a sand garden at its centre and no lawns. This, together with the water feature, reminds us that this is a climate of hot dry summers and droughts. The stark centre is softened by the surrounding regional gardens and historic gardens.

The 20th century garden is more park-like and ornamental, with sweeping lawns, winding paths and feature trees. There is a rose garden and a conservatory for exotic plants.

The 19th century garden had a strong linear design. Trial plantings of native and exotic plants that might be useful for the new colony happened here.

*After the Geelong Botanic Gardens visitor map*

## botanical name *see* scientific name

## bottom heat

Bottom heat is used when propagating plants. It encourages root growth without raising the air temperature around the tops of the plants.

Electrically heated mats or soil-warming cables placed in moist sand with at least 5 cm of sand above and below provide bottom heat. Propagating trays and pots sit on top to gently warm the soil and promote faster germination and root development. A thermostat attached to the device helps to maintain a suitable temperature.

Another method is to make a manure-heated hotbed. A deep layer of decomposing organic matter that will give off heat, like manure and straw, is placed below the soil to warm the surface and air above.

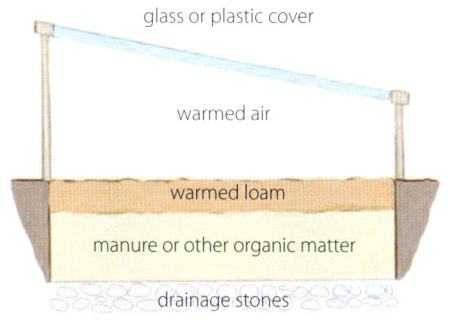

Manure-heated hotbed

Heat mats

---

### bract

A bract is a modified leaf. It is usually positioned at the base of the flower stem.

It may resemble a normal leaf, or it can be colourful, like the bracts of poinsettia (*Euphorbia pulcherrima*) and *Bougainvillea* that surround the tiny inflorescence.

Grasses have papery bracts.

### bracteole

A small bract.

### bramble

A bramble is any plant in the bramble genus *Rubus*. Typically, they have prickly stems called canes and edible fruit, like blackberries, raspberries and boysenberries.

They have a perennial crown and roots, but biennial canes that die after 2 years.

In its first year, the cane is called a primocane. The stem grows and bears leaves, and in autumn it will bear fruit at its tip. In its second year, this same cane is called a floricane. It flowers and fruits in spring, and at the end of the season it dies and is cut back to the crown.

### bridge grafting

Bridge grafting allows nutrients and water to move across the damaged area on a tree trunk.

The wound is cleaned, and scions are placed around the trunk to reconnect the healthy bark above and below the damage.

After the scions have united with the tree, any twigs or leaves arising from them are removed.

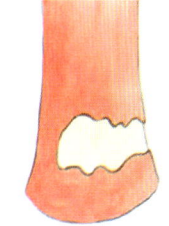

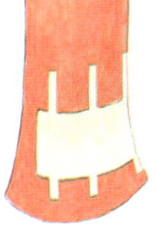

scion

Nutrients and water cannot move across the damaged area on the tree

Scions connect the healthy bark above and below the damage

### brown compost material *see* compost

## bud

A bud is an undeveloped shoot. It can be active or dormant, as those on deciduous plants in winter.

An axillary bud forms in the axil of a leaf. Terminal buds (also called apical buds) are at the tip of a stem, and a lateral bud is on the side of a stem. Stem buds include those on stolons, rhizomes and tubers.

Flower buds grow into a single flower or a cluster of flowers (an inflorescence). These will bear fruit. Leaf buds grow into leafy shoots. Apples and pears have some mixed buds that bear both leaves and flowers, as do stone fruit like cherries, peaches and apricots.

*see also* **spur, sucker, water sprout**

Buds on a potato tuber

## bud union = graft union

## budding

Budding is similar to grafting except that the scion is reduced to a single bud, with a small portion of bark or bark and wood attached.

A small flap of bark is lifted on a suitable rootstock. The scion is placed under this, with the two cambiums aligned, and secured so that the two grow together as one plant. Common methods of budding are T-budding, chip budding and patch budding.

It is commonly used to create a self-pollinating fruit tree by grafting two scions onto the rootstock that will be able to cross-pollinate one another. Budding a scion from a fruit tree onto dwarf rootstock will produce a dwarf fruit tree. Budding is usually done when the bark has started to slip.

BUDDING

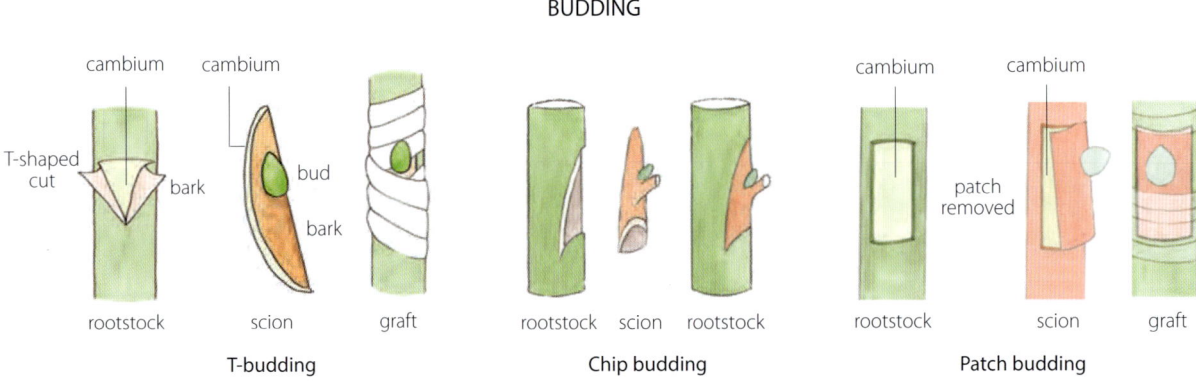

## budwood

Budwood is a piece of stem with vegetative buds that is used in the propagation of new trees.

One bud is taken from the budwood stem for budding. A piece of stem with several buds is taken for grafting. Each is grafted onto a rootstock to create a new tree.

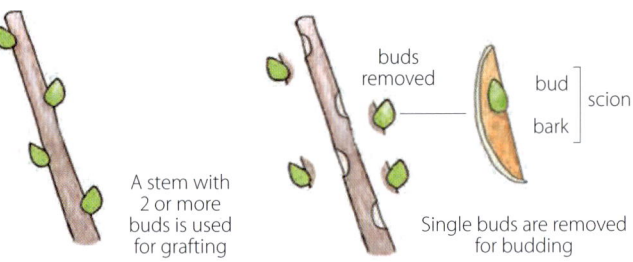

## bulb chipping

Most bulbs can be chipped, but it is time-consuming and can take several years before the bulblets flower.

The bulb is cut vertically, through the basal plate and scales, into sections (chips). The chips are treated with fungicide and put into a planting medium with the basal plate down. The pot with the chips is placed in a plastic bag filled with air and kept in a warm, dark place.

Bulblets will form between the scale leaves just above the basal plate. The scale leaves die. Each bulblet is then potted up.

The bulb is cut vertically through the basal plate

basal plate

chip

Bulblets form on the basal plate between the scale leaves

## bulb scales

Bulb scales on both tunicate and imbricate bulbs are usually fleshy.

They protect the forming flower and store nutrients and water.

## bulb scaling

Healthy scales of a scaly bulb (imbricate bulb) can be detached from the basal plate and propagated.

To propagate, they are cleaned, placed in a plastic bag with a suitable moist medium, and kept in a dark place. Bulblets with roots form after several months and are then potted up. Scales may also break off from the parent bulb in the soil and grow naturally.

leaves

fleshy bulb scales

flower bud

Tunicate bulb

Imbricate bulb

Imbricate bulb

Healthy scales

## bulb sectioning = bulb chipping

## bulbel, bulbil, bulblet

A bulbel is a small bulb. The terms bulbel, bulbil and bulblet are used interchangeably.

Bulbels develop in the axil of a leaf, as tiger lilies (*Lilium lancifolium*), on an above-ground stem, as garlic (*Allium sativum*), or at the base of a mature bulb (an offset), as onion (*Allium cepa*). Bulblets also grown on the underground stems of lilies (*Lilium*). All can develop into a new plant.

*see also* **ferns**

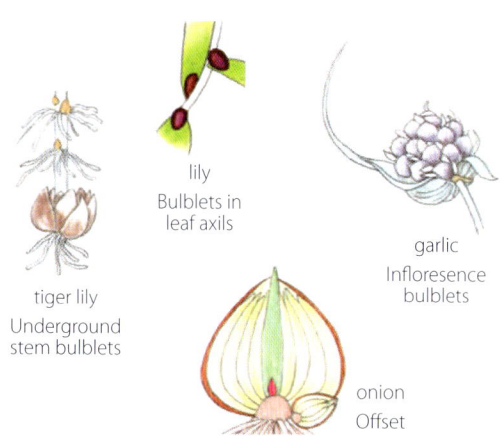

lily
Bulblets in leaf axils

garlic
Infloresence bulblets

tiger lily
Underground stem bulblets

onion
Offset

21

## bulbs

Bulbs contribute to the garden's beauty across the seasons. Daffodils and tulips flower in spring. Gladiolus are summer-flowering. Belladonna lilies and hippeastrums flower from summer through to autumn. Hyacinths can be forced indoors to flower in winter.

A bulb is the resting (dormant) stage of the plant, with water and nutrients stored in its fleshy scales for when it resumes growth.

The fleshy scales enclose an entire plant, with leaves, a flower stem and buds. All of these sit on the basal plate, the bulb's short disc-like stems. The whole structure is usually underground.

There are two main types of bulbs. Tunicate bulbs, like onions and tulips, have a tight-fitting outer coat (tunic) that encloses the fleshy leaf scales, leaves and buds. Scaly (imbricate) bulbs, like lilies, lack a tunic and have free, overlapping fleshy leaf scales. Garlic is an imbricate bulb, with each scale and the entire bulb enclosed in a tunic.

Many bulbs can grow indefinitely and multiply themselves naturally by setting seed or producing bulblets from the base of the parent bulb. Some, like tulips, differ in that the parent bulb dies and is replaced by a new bulb. Most bulb cultivars do not grow true to type from seed and are propagated vegetatively.

Vegetative propagation of tunicate bulbs includes scooping, scoring, twin-scaling, bulb chipping (also called bulb sectioning) and offsets. Scaly bulbs are propagated vegetatively by bulb-scaling and bulblets.

### BULBS

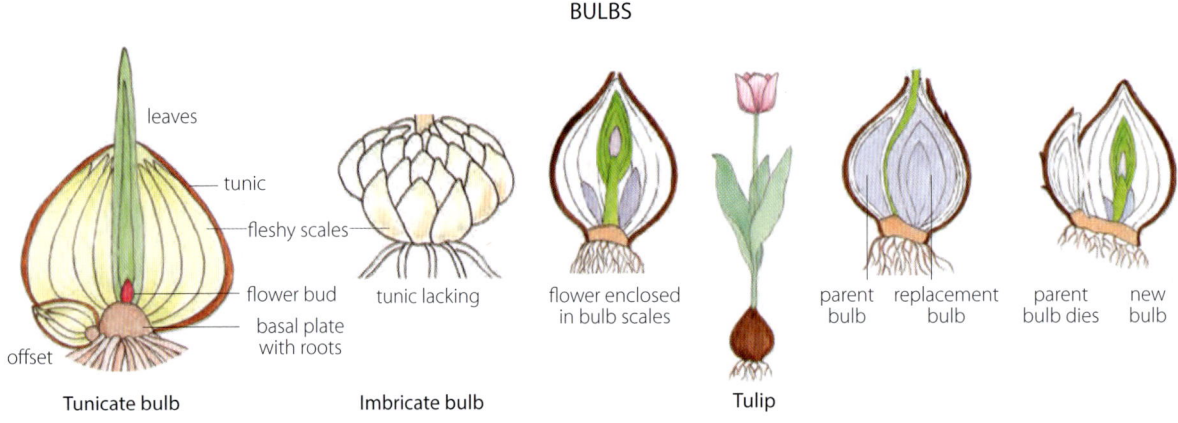

leaves
tunic
fleshy scales
flower bud
basal plate with roots
offset

**Tunicate bulb**

tunic lacking

**Imbricate bulb**

flower enclosed in bulb scales

**Tulip**

parent bulb    replacement bulb

parent bulb dies    new bulb

### VEGETATIVE PROPAGATION OF BULBS

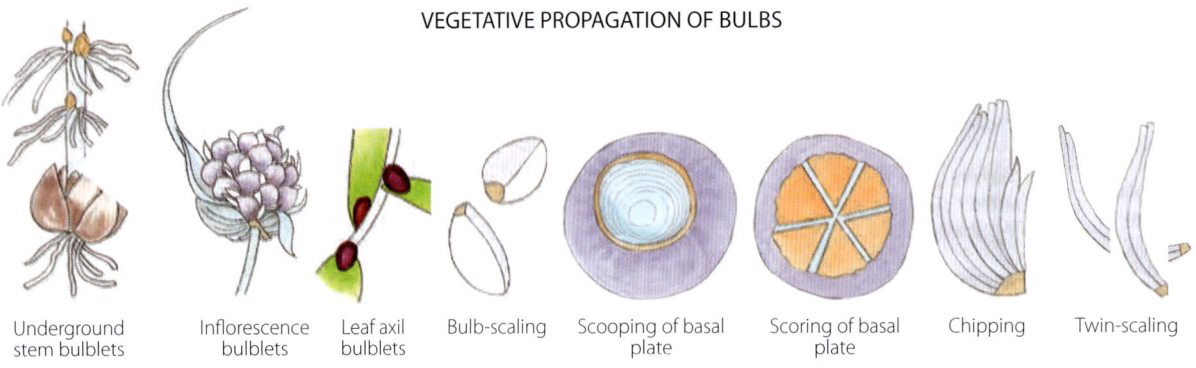

Underground stem bulblets | Inflorescence bulblets | Leaf axil bulblets | Bulb-scaling | Scooping of basal plate | Scoring of basal plate | Chipping | Twin-scaling

**bush** *see* shrub

**bush vs climbing** *see* determinate growth, indeterminate growth

**bush vs vining** *see* determinate growth, indeterminate growth

## buttoning

Buttoning is a condition in which vegetables like broccoli and cauliflower form an unusually small head. It is caused by stress, like temperature extremes or insufficient nitrogen. It is more common among transplanted seedlings.

## cabbage white butterfly *see* biological pest control, horticultural fleece, pests and diseases of gardens

## cactus

Cactus plants are appreciated in Australia as ornamentals. Their stems are succulent and leaves are usually absent. Flowers are mostly solitary and are often large and showy.

All cacti have tiny cushion-like areoles from which branches, flowers and spines arise. Spines are modified leaves that condense moisture from dew and fog and drip it to the shallow roots at the base of the plant.

They are almost entirely native to the Americas, growing naturally in habitats ranging from just south of the Arctic Circle to Patagonia at the southernmost tip of South America, and at altitudes ranging from below sea level (as in Death Valley, California) to over 4800 m high in the Andes. They vary from miniature plants of a few cm to as tall as 20 m.

Cacti reproduce from seeds, offsets, cuttings, division and grafting.

Flowers are usually solitary

Spines grow from areoles

areole

spines

offset

Indian fig
(*Opuntia ficus-indica*)
It lacks spines and is grown for its edible fruit.
In Australia, it is an environmental weed.

globe cactus
(*Mammillaria glochidiata*)

## calcium *see* plant nutrients

## callus

Callus is healing tissue that develops over a wound. It arises from the cambium in wood. In grafting, it promotes the union of the scion with the rootstock.

## callus bridge

The callus bridge develops after grafting, to promote the union of the scion with the rootstock.

It arises from the cambium of both the rootstock and the scion. The cambium produces phloem towards the outside and xylem towards the inside of the callus, to join the rootstock and the scion together.

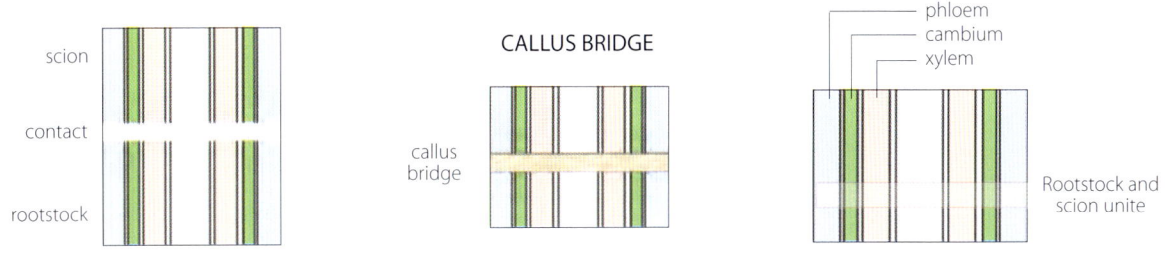

scion

contact

rootstock

CALLUS BRIDGE

callus bridge

phloem
cambium
xylem

Rootstock and scion unite

## calyx *see* flower

## cambium

Of wood, a layer of cells between the bark and the sap wood. It forms new bark cells (phloem) on the outer side and new sapwood cells (xylem) on the inner side.

During grafting, the cambium of the scion and rootstock must be in close contact so that the new bark and sapwood cells made by the cambium can unite the scion and rootstock.

CAMBIUM

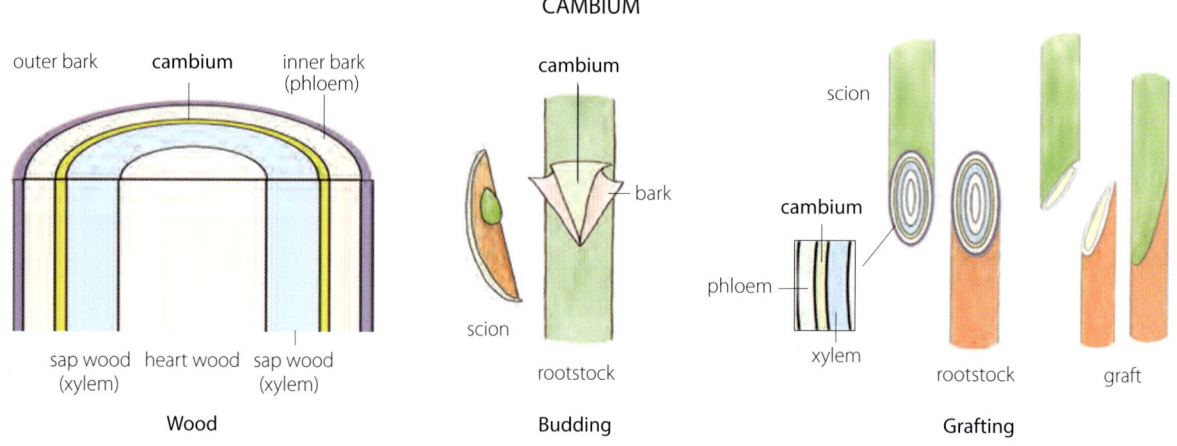

Wood

Budding

Grafting

## cane

One of the usually arching stems that sprout directly from the ground in cane-growing shrubs like weigela (*Weigela*) and mock orange (*Philadelphus*).
A cane is one of the biennial stems of brambles or the prickly stem of a rose. It also refers to the jointed stem of sugar cane or new growth on a grape vine.

## canker

Canker is a serious disease of trees and woody shrubs. It presents as a patch of dead or malformed bark that is a symptom of an injury or a wound being invaded by pathogens.

## catch crops

A catch crop is sown after a main crop has been harvested and the soil is idle before the next crop is planted. It is usually a fast-growing food crop. It will improve soil fertility, control weeds and prevent sun damage that would occur if the soil were bare.

Suitable catch crops are fast-maturing leafy greens like salad leaves, spinach and baby bok choy. All mature within 3 to 4 weeks. In the gap between harvesting brassicas and planting tomatoes, a fast-growing catch crop of clover will add the nitrogen to the soil that tomatoes need.

*see also* **cover crops**

## cauliflory

Having flowers growing from the trunk or older branches of a woody plants, as redbud (*Cercis canadensis*) and its cultivars.

Cauliflory is most common in tropical plants, including cacao (*Theobroma cacao*).

Flowers growing from the branch of a redbud (*Cercis canadensis*)

## chahar bagh garden

In Persian *chahar bagh* means four gardens. It describes symmetrical gardens of four squares or rectangles that came to be called paradise gardens. These gardens were intersected by irrigation channels linked to canals, reservoirs, springs or qanats.

The basic design can be used many times over to make a well-proportioned garden. Mathematical plans were designed to maintain a unified symmetry.

The presence of water and shade, together with the plantings, invited contemplation. There were plantings of fruit trees (apricot, plum, cherry, mulberry, pomegranate and plum), ornamental trees (plane trees, cypress and poplar) and ornamental shrubs (oleander, plumbago and gardenia).

*see* Islamic gardens, Persian gardens

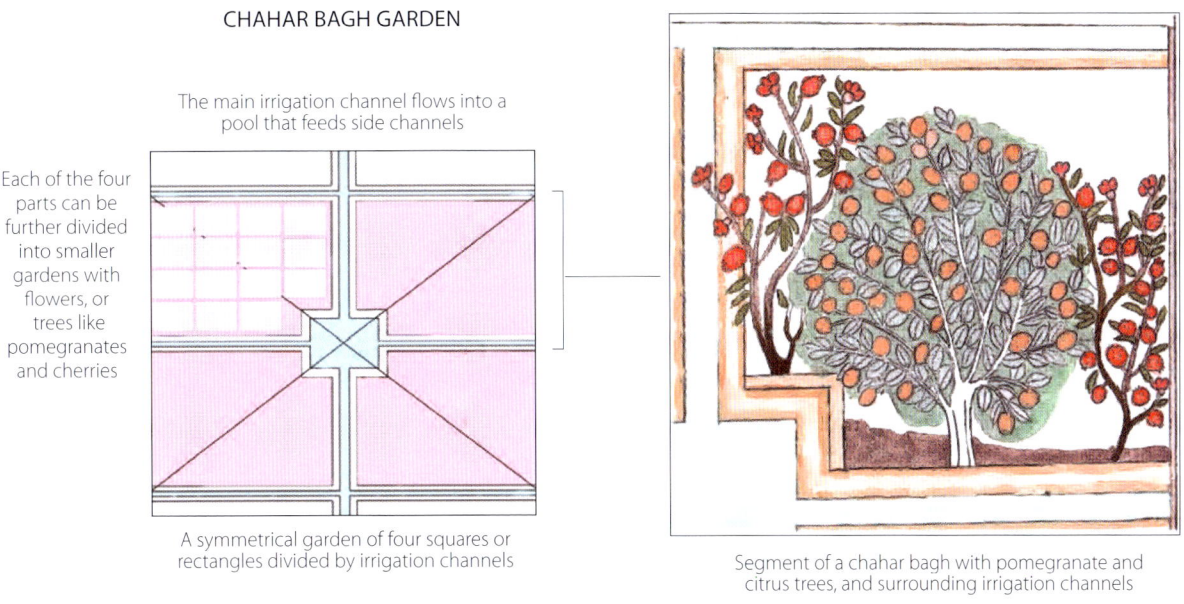

CHAHAR BAGH GARDEN

The main irrigation channel flows into a pool that feeds side channels

Each of the four parts can be further divided into smaller gardens with flowers, or trees like pomegranates and cherries

A symmetrical garden of four squares or rectangles divided by irrigation channels

Segment of a chahar bagh with pomegranate and citrus trees, and surrounding irrigation channels

## chelation

Chelation is a biochemical process that increases the availability of soil nutrients to plants. It occurs naturally in composting, and in soils rich in organic matter and humus. Synthetic chelating agents are available.

## chemical pest control *see* pesticides

## chill factor

The chill factor refers to the number of hours below 7°C (45°F) that some trees need to break dormancy, bloom and bear fruit.

It is divided into three categories: low (up to 450 chilling hours), medium (450–650 chilling hours) and high (more than 650 chilling hours).

Trees that do not get enough chilling hours during the winter may develop flowers at the wrong time of the year or not at all. Breeding methods have created low chill cultivars of fruits, like apples, and nuts that would normally be considered high chill.

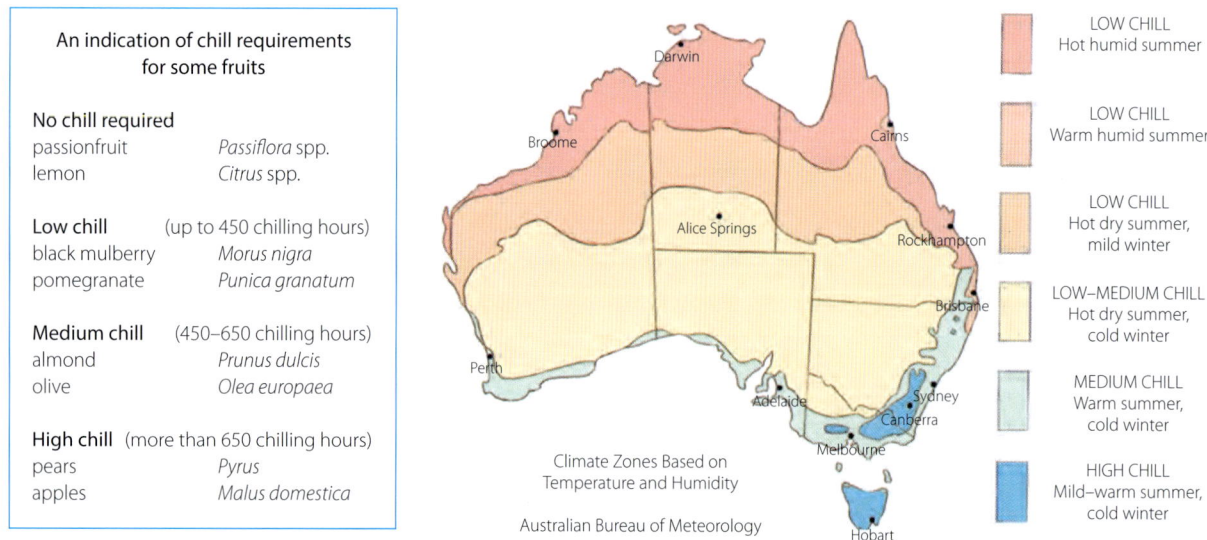

An indication of chill requirements for some fruits

**No chill required**
| | |
|---|---|
| passionfruit | *Passiflora* spp. |
| lemon | *Citrus* spp. |

**Low chill**  (up to 450 chilling hours)
| | |
|---|---|
| black mulberry | *Morus nigra* |
| pomegranate | *Punica granatum* |

**Medium chill**  (450–650 chilling hours)
| | |
|---|---|
| almond | *Prunus dulcis* |
| olive | *Olea europaea* |

**High chill**  (more than 650 chilling hours)
| | |
|---|---|
| pears | *Pyrus* |
| apples | *Malus domestica* |

Climate Zones Based on Temperature and Humidity
Australian Bureau of Meteorology

LOW CHILL
Hot humid summer

LOW CHILL
Warm humid summer

LOW CHILL
Hot dry summer, mild winter

LOW–MEDIUM CHILL
Hot dry summer, cold winter

MEDIUM CHILL
Warm summer, cold winter

HIGH CHILL
Mild–warm summer, cold winter

## chinampas

Chinampas are small artificial islands, built in a shallow freshwater lake, that are used for intensive agriculture. The technique was developed by the Aztecs around their capital Tenochtitlan on the marshes of Lake Texcoco in Mexico. The system is very sophisticated, sustainable and highly productive.

Areas of the lake, about 30–60 cm deep, were chosen to construct rectangular beds about 20 m long and 12 m wide. Willow stakes about 1 m high marked the perimeter, with about 1 to 1.5 m between them. A reed mesh was then woven between the posts to make a fence.

Canals were dug around the garden beds, and the soil and organic matter used to fill the fenced areas to about 50 cm above water level. Willow trees were planted along the borders, their roots anchoring the garden to the lake floor. The chinampas were then ready to be planted out with maize, beans, squash, amaranth, tomatoes, chilli peppers, and flowers. A complex system of drainage ditches, dikes and sluice gates prevented flooding.

The system flourished for about 100 years, from the early 15th century until the Spanish conquest in the 16th century. Tenochtitlan is now the site of Mexico City, and some of the chinampas still exist.

CHINAMPAS

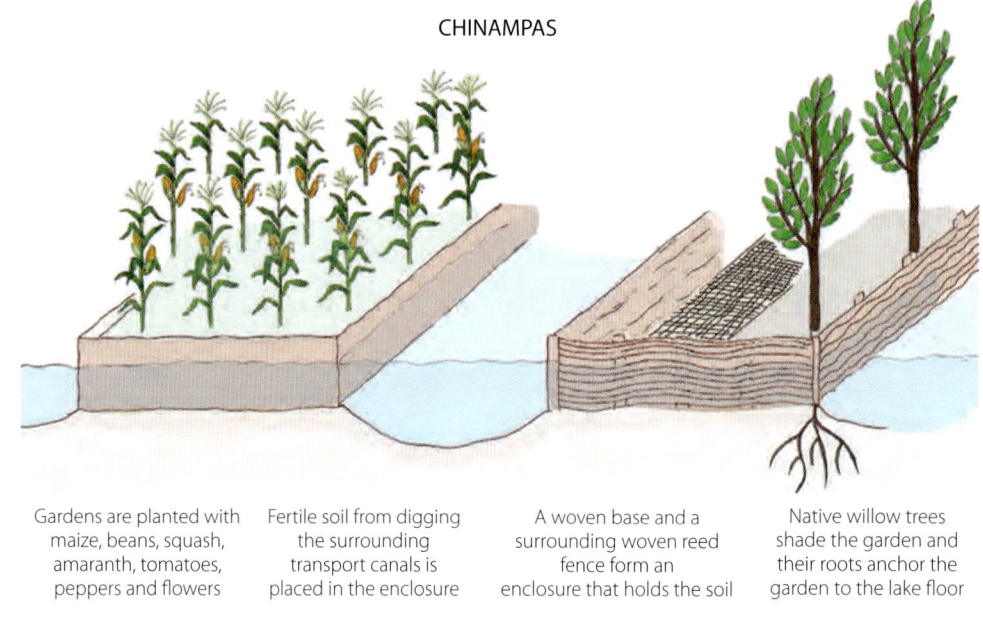

Gardens are planted with maize, beans, squash, amaranth, tomatoes, peppers and flowers

Fertile soil from digging the surrounding transport canals is placed in the enclosure

A woven base and a surrounding woven reed fence form an enclosure that holds the soil

Native willow trees shade the garden and their roots anchor the garden to the lake floor

## Chinese gardens

China has the longest continuous tradition of garden design in the world. Egyptian gardens flowered in the inhospitable landscape of the desert. The Chinese, however, lived in a nature of exceptional beauty and grandeur that inspired a sense of awe and a sense of a person's respectful place within it.

Unlike gardens in the West, the aim was not to bend nature to an ordered, man-made plan, but to create a reflection of the natural world itself. The garden was an intimate expression of the vastness of nature. Features like imitations of hills and meandering watercourses, and carefully placed rocks, reflecting the asymmetry of nature, were present from the earliest times.

Plants were valued for their spiritual and symbolic meaning. The peach promised immortality and the peony wealth and elegance. The garden also engaged the intellect.

There were two great ancient Chinese philosophies. Confucianism taught about human relationships in an orderly society. There were set rules for a person's role and obligations, and the cultivation of conscience and character. Daosim looked beyond society and its rules. It sought meaning in a relationship with the order and harmony of nature. The Chinese gardens invited peaceful contemplation of these philosophies.

A domestic garden was a feature of the household. The dwelling was built around an open courtyard. An outer fenced area might contain a vegetable garden or an orchard. Many kinds of vegetables were grown, including cabbage, turnips, onions, garlic, cucumbers, beans, peas, squashes and melons. Grafting of fruit trees is thought to have been practised around the 1st century BC.

CHINESE GARDENS

*After 'Enjoyment of the Chrysanthemum
Flowers' by Hua Yan, 1753*

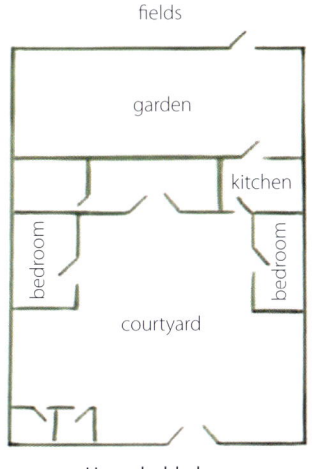

Household plan

**Domestic garden** with vegetables and rows
possibly of fruit trees.
*After 'Vegetable gardeners',
by Shen Zhou, 1496*

## chip budding

Chip budding is used for plants with bark that does not readily separate from the wood.

A triangular chip of bark with a small piece of wood attached is removed from the rootstock. A triangular chip with a bud is then cut in the same way from the plant to be propagated. This is the scion. The scion is inserted in the space made in the rootstock and secured in place with tape.

Once the union is complete, the rootstock is cut back to just above the new shoot.

**CHIP BUDDING**

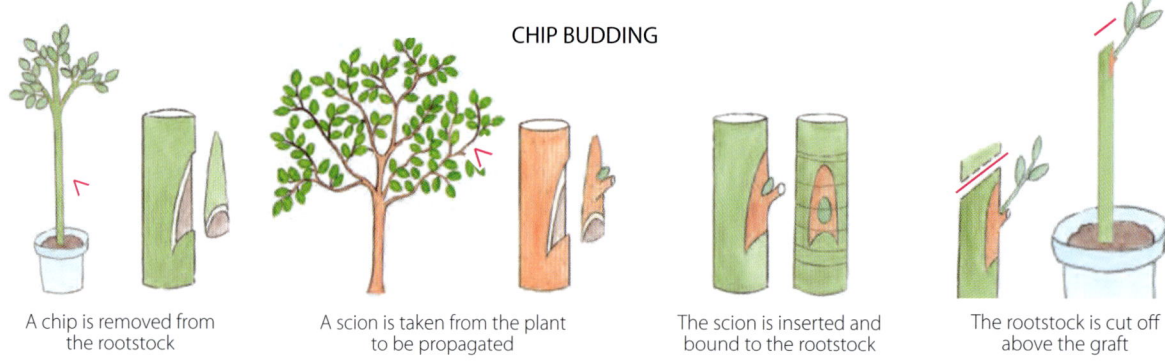

A chip is removed from the rootstock

A scion is taken from the plant to be propagated

The scion is inserted and bound to the rootstock

The rootstock is cut off above the graft

## chitting

Chitting is the practice of exposing seed potatoes to light to encourage them to shoot before being planted.

Larger seed potatoes can then be cut into pieces, each with shoots. Both the shooting seed potatoes and cut pieces will form clones of the parent plant.

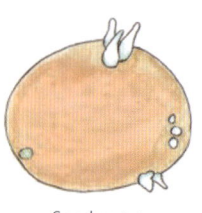

Seed potato

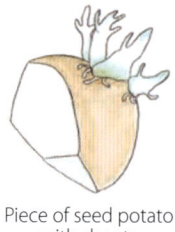

Piece of seed potato with shoots

## chloride *see* plant nutrients

## chlorophyll

A greenish pigment found in plants. Chlorophyll absorbs light energy from the sun for photosynthesis.

## chlorosis

Chlorosis is an abnormal condition resulting in green plants becoming yellowish due to a reduction in chlorophyll levels. Its causes include mineral deficiency and disease. It is particularly noticeable in leaves.

Leaf chlorotic between the veins

blueberry (*Vaccinium corymbosum*)

## chop and drop

Chop and drop refers to cutting down a plant at the base, chopping it into pieces, and laying them on the soil to rot down. The material should be processed before it sets seed.

A thin layer of manure topped with a layer of mulch can be added. The process is done with vegetables that have finished, some prunings, and green manure crops, like buckwheat and clover. The availability of water will affect the rate at which the material decomposes.

Chop and drop doesn't work well in dry climates.

## citrus

Citrus are aromatic, evergreen shrubs or small trees belonging to the genus *Citrus,* in the rue family Rutaceae.

Citrus includes lemons, oranges, limes and grapefruit, all of which are hybrids. Natural citrus fruits include mandarin, pomelo and citron.

The fruit is a segmented berry (hesperidium) with a leathery skin and juicy sacs inside the inner segments. The leaves and outer skin of the fruit are covered in aromatic oil glands.

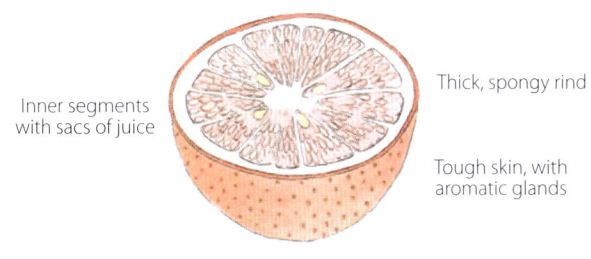

Inner segments with sacs of juice

Thick, spongy rind

Tough skin, with aromatic glands

sweet orange (*Citrus sinensis*)

The sweet orange (*Citrus sinensis*) is a hybrid between a pomelo (*Citrus maxima*) and a mandarin (*Citrus reticulata*)

## clay

Clay is made up of particles that are shaped like flakes. It retains nutrients, but in the winter it is wet and sticky, and in the summer it bakes hard and dry. A heavy soil has a high proportion of clay and is difficult for plant roots to penetrate. Sand particles range in size from 0.05–2.0 mm, silt particles from 0.002–0.5 mm and clay particles are less than 0.002 mm.

## clayey soil

Heavy clayey soils can be improved by aerating the soil and regularly adding organic matter and compost. Sand will further compact clay and gypsum will break down some, but not all, clay soils. Cover crops like comfrey and some daikon radish cultivars have roots that penetrate and loosen this kind of soil. A garden bed can be established on top of clayey soil using sheet composting.

## clean cultivation

In clean cultivation, the space between plants in a garden is kept clear by tillage and the removal of weeds.

## cleft grafting = wedge grafting

## climate zones *see* next page

## climbers

Climbers have vigorous shoots that use plants, trees and other structures for support. They include twiners, vines and bines.

Climber species use different methods to climb. English ivy has hairs along its stems that secrete a glue-like substance that adheres it to a surface. Climbing and rambling roses need to be tied to a support. Wisteria and hop vines have stems that twine around a support, and grape vines use tendrils to attach themselves to a support.

twining stem

English ivy    adhesive roots

hops

tendril

grapevine

## climbing roses

Climbing roses are less vigorous than rambling roses. They are shrubs with stiff, arching stems that can be trained over a trellis or to cover a wall.

*see also* **roses**

## climate zones

The Australian Bureau of Meteorology classifies Australia into six major climate zones: Equatorial, Tropical, Subtropical, Desert, Grassland and Temperate. Temperate zones are often further divided into Warm Temperate, Mediterranean, Cool Temperate and Alpine.

The zone you live in will help determine what will grow in your garden. With climate change, intensive plant breeding programmes are underway to develop varieties that will thrive in different zones, making the choices of varieties for planting much larger. For example, low chill apple varieties (100–300 hours) have been bred to suit tropical, subtropical and warm temperate conditions.

### CLIMATE ZONES

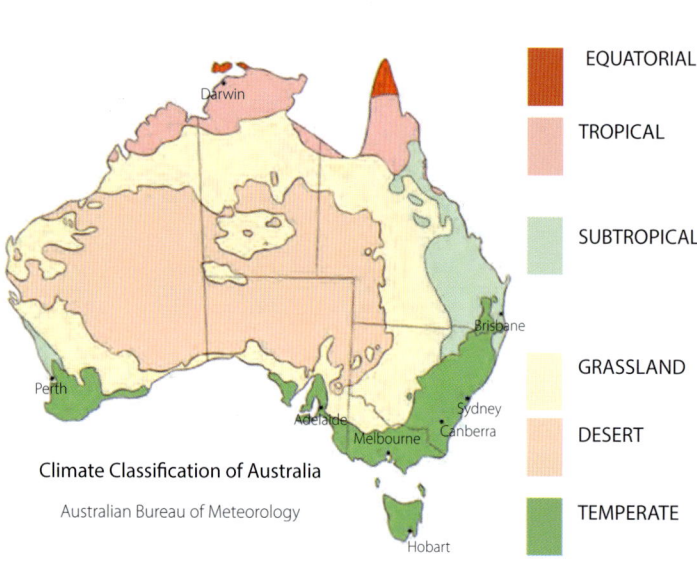

Climate Classification of Australia

Australian Bureau of Meteorology

**EQUATORIAL**

**TROPICAL**

**SUBTROPICAL**

**GRASSLAND**

**DESERT**

**TEMPERATE**

**Tropical and Equatorial zones** have two seasons: a hot, humid wet season (November to April), with January temperatures higher than 30°C (86°F), and a dry season. Tropical food gardens have mostly perennial plants like sweet potato, yams and pawpaw.

The **Subtropical zone** has a wet, warm and humid summer. January temperatures are below 30°C (86°F). Gardens have perennial plants like sweet potato, yams and pawpaw.
A large variety of annuals are also grown. Temperate zone annuals can be grown before the hot summer arrives.

**Grassland and Desert zones** have very high summer temperatures and prolonged droughts are common. Some annuals can be sown in Autumn.

**Temperate zones** experience four seasons.
Warm temperate areas, like the coastal regions near Sydney, have moderate winters.
Mediterranean areas, like much of coastal South Australia, have rainfall in winter and warm summers.
Cool temperate areas, like Canberra and Tasmania, have cold winters with frost and a short summer growing season.

---

## cloche

A cloche is a bell-shaped cover used to protect single plants from weather and pests.

Traditionally, they were made from glass, but are now available in plastic, chicken wire and bamboo. Empty plastic containers can be modified to make a cloche.

## clone

A clone is produced by vegetative propagation rather than from seeds. It is genetically identical to its parent.

## cloud pruning

A Japanese method of pruning, *niwaki*, that creates trees with outstretched branches and rounded canopies of foliage resembling clouds. *Niwaki* is a garden tree, as distinct from *bonsai*, that is a miniature potted tree.

This sculptured style reveals the structure and essence of the tree and keeps its size in harmony with the rest of the garden. Trees are usually asymmetrical but beautifully balanced at the same time. Shrubs are also trimmed and pruned into rounded cloud-like shapes. Modern gardens of all styles have adopted this as a more relaxed version of topiary.

Cloud pruned azaleas

## clump

A clumping plant grows outward from the centre and has a compact shape. Daylilies, bromeliads and mondo grass are examples.

Clumps can be divided with a spade or sharp knife for propagation.

A clump can also be made artificially. Clumping birches are created by planting a single-stemmed tree and, after a season or two, cutting the trunk back to a few centimetres above the ground. Of the numerous shoots that come up from the stump, 3–5 are retained to become trunks.

Another way of achieving a clump is to plant 3–5 trees in a single hole, with the stems spaced about 30 cm apart.

*see also* **bamboos, coppicing**

Clumps can be divided for propagation

grass-leaved scabiosa (*Scabiosa graminifolia*)

clumping bamboo

**cobalt** *see* plant nutrients

## coco peat

A soft, spongy medium that is used in potting mixes and some soils.

It helps hold water in sandy soils and to loosen the texture of clay soils. It can be moulded to make liners for hanging baskets, pots for plants, and trays for raising seedlings. Coco peat degrades in the soil, and the pot, plant and medium can all be planted directly into the garden.

Also called coir, it comes from the fibrous layer surrounding the inner hard shell of a coconut. It is valued as an alternative to sphagnum moss and peat.

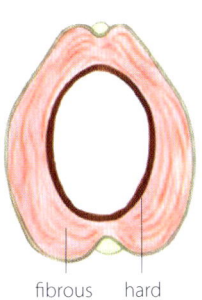

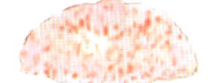

fibrous layer  hard shell

Coconut

The pot, plant and medium can all be planted directly into the garden

Garden and potting coco peat

31

**codling moth** *see* pests and diseases of gardens

**coir**

Coir comes from coconuts. The outer fibrous layer surrounding the inner hard shell is shredded, ground or cut into mulch chips and sold as coir.

Also called coco peat, it is a soft spongy medium that is used in potting mixes, to help loosen the texture of clay soils, and to hold water in sandy soils. It can be moulded to make liners for hanging baskets, pots for plants, and trays for raising seedlings.

Coir degrades in the soil. The coir pot, plant and potting mix can all be planted directly into the garden.

Coir is sold in dry compact blocks that expand when soaked in water. It has become valued as an alternative to peat moss, perlite and vermiculite.

COIR

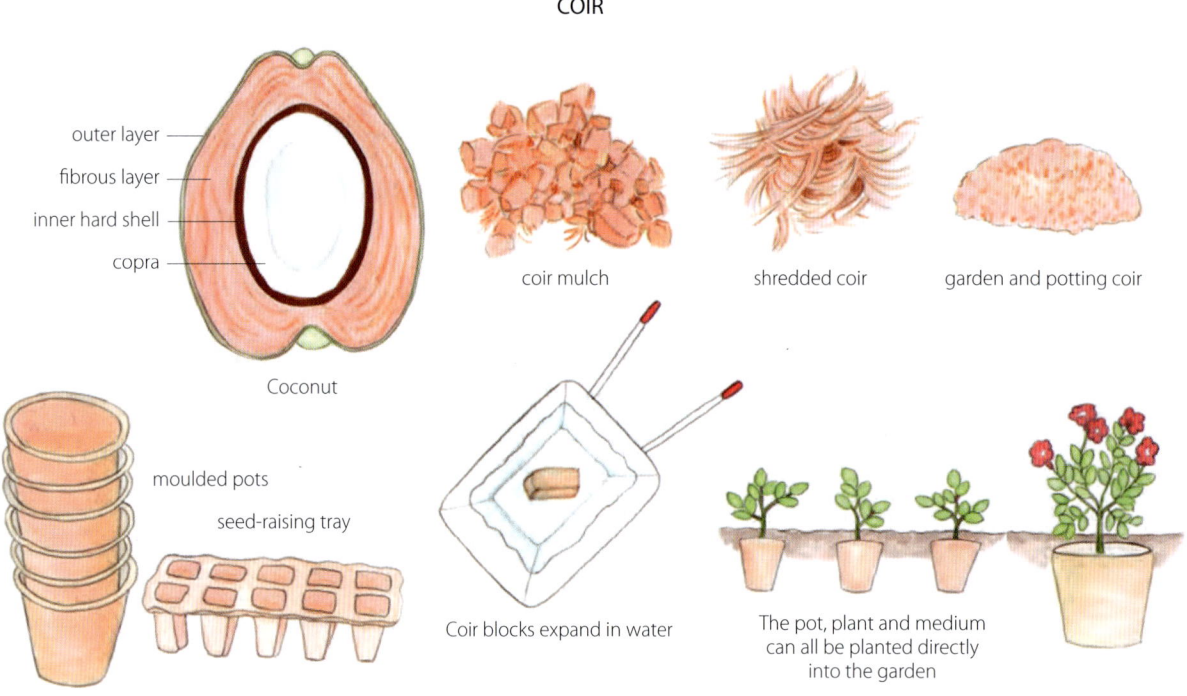

outer layer
fibrous layer
inner hard shell
copra

Coconut

coir mulch

shredded coir

garden and potting coir

moulded pots

seed-raising tray

Coir blocks expand in water

The pot, plant and medium can all be planted directly into the garden

**cold composting** *see* compost

**cold frame**

A cold frame is an unheated rectangular structure, with sides made of wood or bricks, and a glass top that can be propped open for ventilation.

It provides natural sunlight and an extended period of warmth at the beginning of autumn and spring.

A cold frame is commonly used to overwinter semi-hardy plants, to grow seedlings and to harden off young plants.

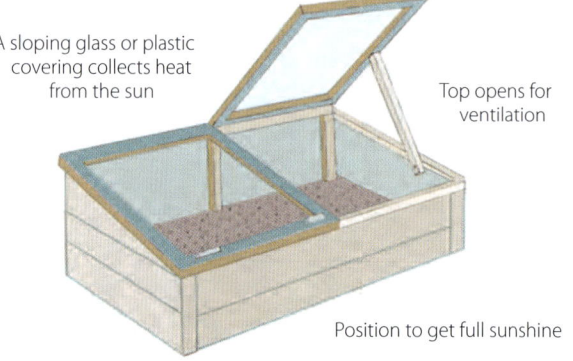

A sloping glass or plastic covering collects heat from the sun

Top opens for ventilation

Position to get full sunshine

## common name

A common name is the local or popular name for a plant, as opposed to the scientific name. Hawthorn, single-seeded hawthorn, may blossom, maythorn, quickthorn and whitethorn are all common names for *Crataegus monogyna*.

## companion planting

Companion plants in the garden help each other grow rather than competing against one another.

They can attract beneficial insects, repel pests, provide nutrients, and give physical support or shade for other plants. There are theories that plants can chemically enhance or inhibit each other's growth and well-being.

Gertrud Franck observed plants over many years and noted those that benefitted each other when they grew together, and those that were incompatible. In 1983, she published her ideas in her now classic book *Companion Planting: Successful gardening the organic way* .

### Plants that make good neighbours

| | |
|---|---|
| beans / brassicas | lettuce / kidney beans |
| brassicas / beetroot | lettuce / beetroot |
| tomatoes / parsley | lettuce / marigolds |
| tomatoes / onions | peas / brassica |
| tomatoes / celeriac | celery / all types of greens |
| carrots / onions | cucumbers / brassicas |
| lettuce / radish | potatoes / all types of brassicas |
| parsnips / onion | potatoes / peas |
| lettuce / cucumbers | potatoes / broad beans |

### There are a few unfavourable combinations

| | |
|---|---|
| beans / onions | red cabbage / tomatoes |
| cabbages / onions | parsley / cabbage, lettuce |
| betroot / tomatoes | potatoes / onions |

Gertrud Franck *Companion Planting: Successful gardening the organic way*

## complete fertiliser

A complete fertiliser is any commercial fertiliser with all three nutrients, nitrogen (N), phosphorus (P) and potassium (K), present. The percentage of each is printed on the bag or container. Other soil nutrients may also be included.

## complete flower

A complete flower has all four whorls of sepals, petals, stamens and pistils.

Complete flower

sepal
petal
stamen
pistil

## compost see page 34

## compost accelerator see compost starters

## compost bin

Compost bins are good for gardeners with limited space. They are a tidy way of composting some kitchen scraps and garden waste.

You can choose to have an open or lidded bin. A bin with a lid can speed up composting, keep out rodents, and prevent the compost from getting too wet when it rains.

It is tricky to aerate these bins. Without aeration it will take up to a year to make compost.

Closed compost bin

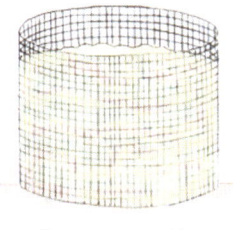

Open compost bin

## compost booster see compost starters

## compost

Compost is naturally decomposed organic matter, such as leaves, lawn clippings, eggshells and manure.

Over time, organisms like bacteria, fungi, nematodes and worms completely break down the organic matter into what looks like a dark, odourless soil that is rich in nutrients. The compost is spread over the garden and topped with a layer of protective mulch.

Organisms that decompose organic waste need materials that contain nitrogen (greens) and carbon (browns), as well as air and water to thrive. The right combination of these will yield the best compost. Turning the compost aerates it and adds oxygen. Temperature will affect the rate at which the compost forms.

### WHAT TO PUT IN YOUR COMPOST

Nitrogen (greens)  
Carbon (browns)  
Oxygen (air)  
Water  

1 part greens  
2 parts browns  
Turn compost to aerate it  

Compost is built up in layers of green and brown material

| 'Greens' for nitrogen | 'Browns' for carbon | DO NOT add |
|---|---|---|
| vegetable and fruit scraps | leaves | meat, fish, poultry |
| coffee grounds and tea bags | shredded straw, hay | bones |
| fresh grass clippings | sawdust | milk products |
| chicken/livestock manure | ashes from wood | pet manures |
| green plant cuttings | shredded paper (not glossy) | diseased plants |
| | | weed seeds |

There are two main types of composting: cold composting, also called passive composting, and hot composting, also called active composting.

In cold composting, materials are placed in a heap, bin, or in layers on the ground (sheet composting) and left until they break down slowly over 6–12 months or more.

Hot composting is much quicker than cold composting. It also has the benefit of killing weed seeds and pathogens that cause disease. The size of the bin is important for maintaining heat. It should be about 90 cm wide, deep and high. Active microbes using oxygen cause heat to build up in the compost. When the pile reaches 70°C, it is turned to aerate it and introduce more oxygen. It results in a fine material. Examples of hot composting methods include the three-bin composting method and tumbler bins.

In worm farming, worms produce nutrient-rich castings that can be added to the soil.

Some composting, called anaerobic composting, is done without oxygen. The material is buried in the ground and allowed to break down over time. Examples are trench and pit composting, and the bokashi method.

## compost problems

Most compost problems can be rectified by getting the balance of the ingredients correct. Microorganisms, for example, can't work in a pile that has too much or too little water. There are also commercial compost starters available.

### IF COMPOSTING IS NOT WORKING

| SYMPTOMS | PROBLEM | START THE COMPOSTING PROCESS |
|---|---|---|
| Material is moist and dense but not wet | The material is not decomposing | The material is lacking oxygen. Fluff or turn the pile with a pitchfork to aerate it. |
| Material is wet, smelly and may be oozing | The material is saturated with water | Turn the pile and add dry brown material such as chopped straw and sawdust. Cover the pile to protect from rain. |
| Material is dry to the touch. There is very little living activity in the pile. | Dry browns are not breaking down | The material lacks water. Add greens such as lawn clippings and kitchen scraps. Turn and moisten the pile throughout. |
| The microorganisms are not working efficiently | The material is not heating up (with hot composting). It is slow to decompose. | The material lacks nitrogen. To restore the carbon–nitrogen balance, add well-rotted chicken manure and grass clippings. Sprinkle high-nitrogen material, like blood and bone, over the compost. |
| The material is giving off a smell like ammonia | The material has an unpleasant odour | The material has an excess of nitrogen caused by adding too much green material. Add brown materials like sawdust, chopped straw and shredded paper or cardboard to restore the carbon to nitrogen balance |

## compost starter

A compost starter is put in your compost pile to activate it and make it break down more quickly.

*see* compost problems

### compost tumbler

A compost tumbler is a fully sealed bin or drum that is rotated to mix the materials inside. Microorganisms breaking down the compost create heat that decomposes the contents.

Chopped green and brown composting material is added to the bin, preferably in a single batch, in a ratio of 1 part green to 2 parts brown. For the first few batches, a compost activator, such as well-rotted chicken or cow manure, will get the process started more quickly. The tumbler is rotated regularly to mix and aerate the contents.

A steady moisture level and a temperature of over 60°C are crucial to kill weed seeds. If worms are added, a hot environment will kill them.

Compost will take longer to form at lower temperatures.

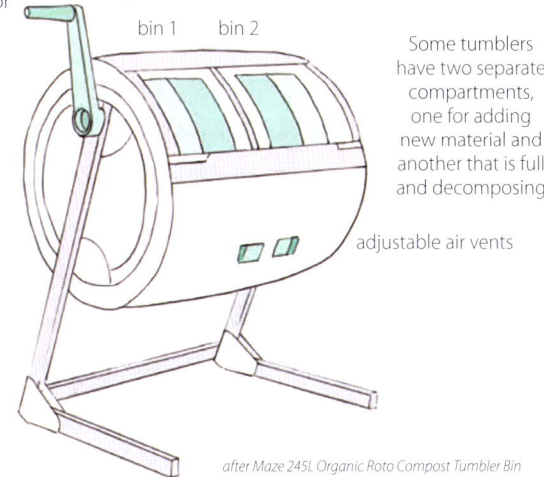

handle for rotating the bin

bin 1   bin 2

Some tumblers have two separate compartments, one for adding new material and another that is full and decomposing

adjustable air vents

*after Maze 245L Organic Roto Compost Tumbler Bin*

**Green materials include**
vegetable and fruit scraps
coffee grounds, tea bags
fresh grass clippings
chicken/livestock manure
green plant cuttings

**Brown materials include**
leaves
shredded straw, hay
sawdust
ashes from wood
shredded paper

## composting worms

Composting worms are different from the earthworms found in garden soil. They are surface feeders and won't burrow into garden soil. They prefer wetter conditions than earthworms and need a thick layer of organic material to feed on.

There are various breeds of composting worms, the most common being the tiger worm that consumes its own body weight in a day and multiplies quickly when well fed.

*see also* **earthworms, worm farming**

## compound layering = serpentine layering

## copper *see* plant nutrients

## coppicing

Coppicing involves cutting a tree or shrub down to ground level and allowing the living stump to generate new shoots for years to come.

A coppice creates dense greenery and is maintained by pruning it back at regular intervals.

This method has been used for hundreds of years to provide wood for fuel, hedging, wind breaks, and thin stems, as from a coppiced willow, for crafts and fences. It is now used for an ornamental effect and for maintaining long-term attractive juvenile growth.

Not all plants are suitable for coppicing.

shoots

stump

## cordon fruit trees

Cordon fruit trees are grown as a single spur-bearing stem. This results in very compact side branches and a good fruit harvest for the size of the plant and the space taken.

Suitable spur-bearing trees include apples, pears, cherries and plums. They are pruned each year to maintain their columnar shape.

Several vertical cordons can be grown in a row, making them ideal for small garden spaces. They are best placed against a wall, fence or trellis for support.

Grape vines can be pruned to have horizontal cordons.

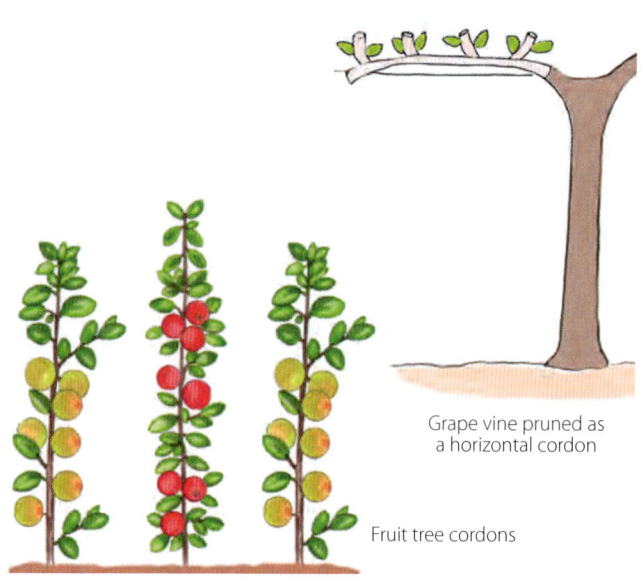
Grape vine pruned as a horizontal cordon

Fruit tree cordons

## corm

A corm is a solid, rounded underground stem base that stores food. The centre sits on the basal plate and is surrounded by dry scale leaves that form a protective tunic.

Corms are annual and are replaced each year with a new corm. Crocus (*Crocus*) and gladioli (*Gladioli*) are examples. They produces a new corm on top of the old corm. as well as cormels around the base of the mother corm. Cormels can be propagated. Bananas have corms, but are propagated by suckers.

Cormels of taro (*Colacasia esculenta*) are used for food and propagation.

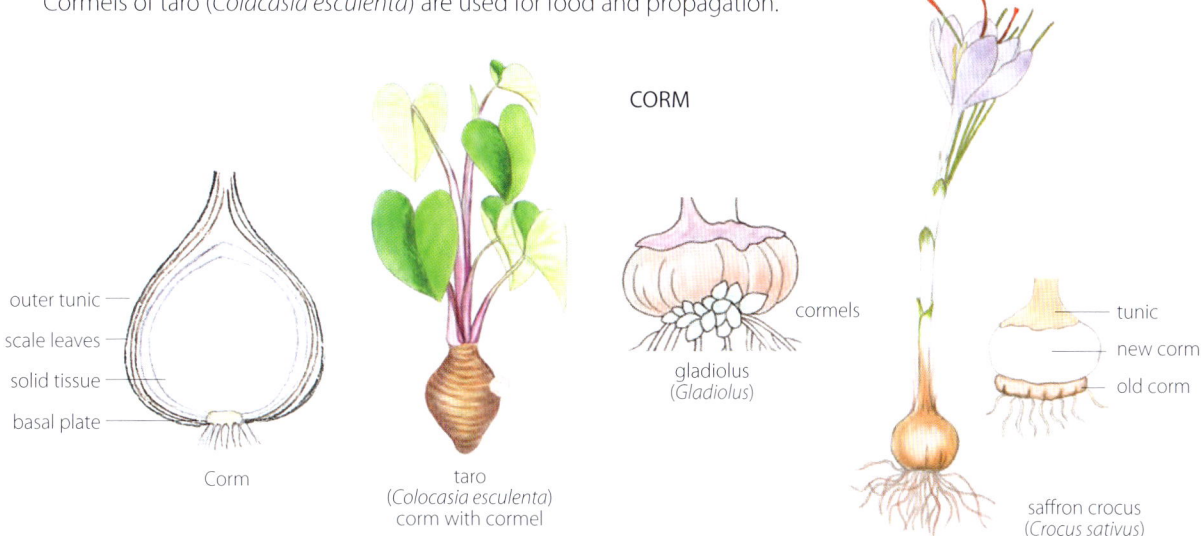

CORM

outer tunic
scale leaves
solid tissue
basal plate

Corm

taro
(*Colocasia esculenta*)
corm with cormel

cormels

gladiolus
(*Gladiolus*)

tunic
new corm
old corm

saffron crocus
(*Crocus sativus*)

## cormel

A cormel is a small corm developed at the base of the mother corm.

It can be detached and used for propagation.

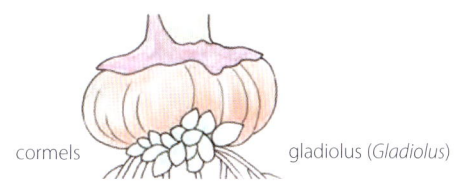

cormels          gladiolus (*Gladiolus*)

## corolla *see* flower

## cottage garden

A small garden, traditionally surrounding a cottage, densely planted with flowers herbs and vegetables. The look is an informal jumble with plants mixed together rather than in neat rows. It may have fruit trees or a bee hive. Despite its appearance, it needs considerable care to prevent it becoming unruly and weedy.

The cottage garden originated in England towards the end of the 19th century and remains popular today.

Favourite cottage plants include roses, lupins, foxgloves, aquilegias, honeysuckle and hollyhocks

## cotyledon

A cotyledon is one of the leaves inside a seed that has a supply of nutrients for use during germination. Cotyledons are also called seed leaves.

Cotyledons of seeds that germinate underground remain in the seed coat, and the first leaves to appear above ground are true leaves. Cotyledons of seeds that germinate above ground grow upwards and are the first leaves to appear. They are followed by true leaves.

True leaves look like the leaves of a mature plant and can photosynthesise. They take over the nourishment of the plant, and both above-ground and underground cotyledons then wither away.

A seed can have a varying number of cotyledons. Flowering plants (angiosperms) with two cotyledons are mostly eudicots. Monocotyledons are flowering plants with a single cotyledon. Non-flowering seed plants (gymnosperms) have two or more cotyledons.

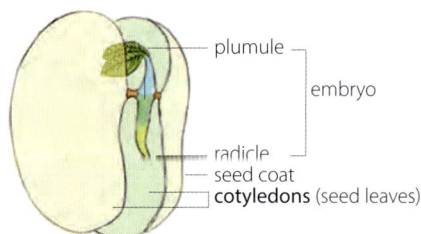

Broad beans (*Vicia faba*) have two cotyledons

## cover crops

Cover crops are temporary, fast-growing plants grown on soil that would otherwise be bare. They prevent erosion from heavy rains and wind, stop weed growth, reduce compaction and add organic matter that enriches the soil. As they die down they release valuable nitrogen, phosphorous and sulphur back into the soil.

A cover crop can be seeded in one bed or over entire sections of a garden. They include buckwheat and nitrogen-fixing plants like beans, alfalfa, clover, lupins and peas. Clover and alfalfa have extensive root systems that rot and become organic matter. Oats and radishes trap free nitrogen in the soil that would be lost through leaching. The crop is cut back before seed heads form and left to break down on top of the soil as a mulch or dug in.

Cover crops are used in crop rotation. Knowing what vegetables will be planted next determines the type of cover crop used.

*see also* **catch crops**

## creepers

A creeper grows horizontally and sends out roots from nodes on its stem as it spreads. Creeping thyme (*Thymus serpyllum*) can be used as a ground cover in sunny sites. Creepers can work well as a filler in rock gardens and between stepping stones.

## Critical Root Zone *see* drip line

## crop rotation

The crop rotation system alternates different crops in the same garden bed, traditionally over three or four years. A four-year rotation would use four beds, allowing the same crop to be grown in the same bed every four years.

Each type of plant takes up specific nutrients from the soil and returns them as benefits for the next crop. Legumes, like peas, beans and alfalfa, fix nitrogen from the air that is necessary for healthy leafy vegetables like lettuce, cabbages and spinach. Brassicas, like cabbage and cauliflower, break up the soil and have a toxic effect on some soil-borne disease organisms. Root vegetables draw on nutrients from layers deeper in the soil than other vegetables.

Some systems do crop rotation according to vegetable families. Another does rotation according to whether the plants are legumes, root vegetables, leafy greens or fruiting vegetables. Systems are being developed where there may be six or more rotations. Animal feed like grasses and clover can also be included in the sequences.

Succession planting, intercropping and companion planting can work well with crop rotation.

*cf.* **ecological gardening**

**LEGUMES**

(Add nitrogen to soil)

Peas          Soy beans
Green beans   Chickpeas
Broad beans   Alfalfa
Lima beans

**ROOT VEGETABLES**

(Light feeders from deeper soil)

Carrot    Leek
Radish    Parsnip
Onion     Turnip
Garlic    Beet

**GREENS AND BRASSICAS**

(Require nitrogen)

Lettuce           Bok choy
Spinach           Broccoli
Cabbage           Cauliflower
Brussels sprouts  Herbs

**FRUITING VEGETABLES**

(Require phosphorus)

Tomato    Eggplant
Pumpkin   Cucumber
Melons    Potatoes
Peppers   Corn

**cross-fertilisation** *see* fertilisation

## cross-pollination

Cross-pollination is the transfer of pollen from the flower or cone of one plant to the egg-bearing flower or cone on different individuals of the same species or of a different variety of the same species.

It is found in both angiosperms (flowering plants) and gymnosperms (cone-bearing plants). The movement of pollen may occur by wind, as in conifers, or by insects and birds in flowering plants. In angiosperms, it can also be done artificially by hand.

**crown** *see* next page

## crumb

Crumb refers to the way sand, silt and clay are bound together in clumps. The crumb varies in size and has air pockets throughout called pores. Soil pores (porosity) determine how well air and water move through the soil.

*see also* **soil aggregates**

## crown

The crown of a tree is made up of the branches and foliage above the trunk.

The cluster of fronds at the tip of a palm stem is the crown.

The crown of shrubs, perennials and annuals is the tissue that connects the stems with the roots. Leaves, flowering stems, rhizomes, stolons and grass tillers arise from the crown.

The crown of many plants, like ferns and asparagus, can be divided for propagation.

CROWN

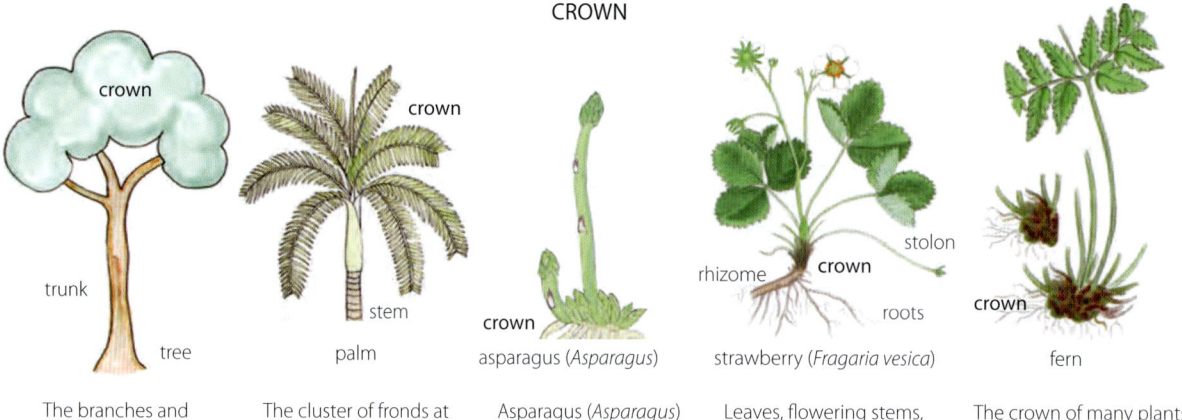

| The branches and foliage of a tree above the trunk are the crown | The cluster of fronds at the tip of a palm stem is the crown | Asparagus (*Asparagus*) spears are new shoots that grow from the crown | Leaves, flowering stems, stolons and rhizomes arise from the crown | The crown of many plants can be divided for propagation |

## cultivar

Cultivars are plants that have been bred by humans, usually for specific traits such as taste, colour, or resistance to disease.

Plants that can be considered to be cultivars include deliberate hybrids, accidental hybrids, and plants resulting from selective breeding. Plants, like apples and broccoli, have many cultivars.

The term 'cultivar' derives from 'cultivated variety'. A plant that is a cultivar is indicated by the cultivar name coming after the genus and species names, as the apple cultivar *Malus domestica* 'Granny Smith'.

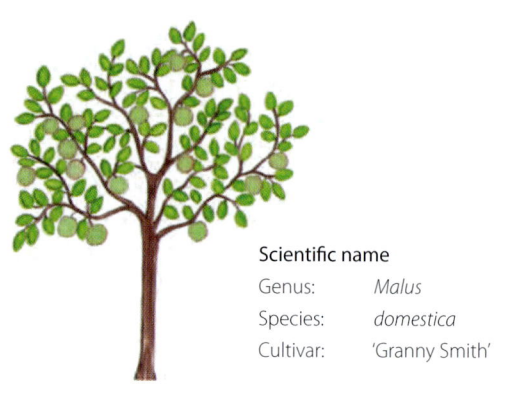

**Scientific name**
Genus:     *Malus*
Species:   *domestica*
Cultivar:   'Granny Smith'

**Common name:**   Granny Smith apple

## cut and come again

Cut and come again refers to harvesting only the outer leaves of a plant and allowing the rest to continue to grow.

This provides a continuous harvest of leaf from leafy vegetables and herbs, like rocket, basil, parsley, bok choy, spinach and leaf lettuce.

## cutting back *see* heading back

## cuttings

A cutting is a piece of stem, leaf or root, that is encouraged to form roots and shoots of its own in order to produce a new plant. The new plant is identical to its parent.

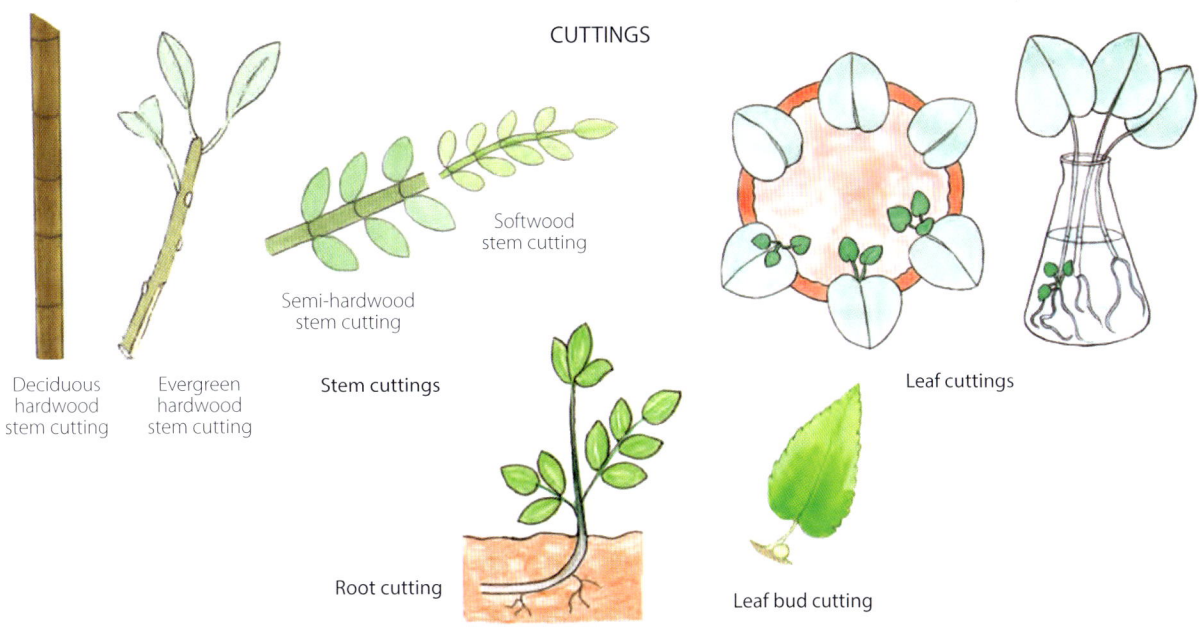

CUTTINGS

Softwood
stem cutting

Semi-hardwood
stem cutting

Deciduous
hardwood
stem cutting

Evergreen
hardwood
stem cutting

Stem cuttings

Leaf cuttings

Root cutting

Leaf bud cutting

## cycads

Cycads are ancient plants that haven't changed much since they were plentiful in the time of the dinosaurs more than 200 million years ago.

They are unlike any other living plant and are not palms, despite their common name of sago palm. Many are grown as ornamentals. The trunk has a rosette of divided leaves on top and is usually above ground. Male and female reproductive cones are on separate plants.

Their usually colourful seeds are unique in that they do not go into dormancy like other seeds. They dry out quickly and die if not planted straight away. Some cycads can be propagated from offsets at the base of the parent plant.

Leaves are divided and palm-like

The male or female cone is produced among the leaves from the pithy crown at the top of the trunk

Cycads are long-lived and most are slow-growing

**damping off** *see* pests and diseases of gardens

**dams** *see* next page

## deadheading

Deadheading is the removal of fading or dead flowers to prevent them going to seed. It is usually done with scissors or the fingers and thumb. This encourages the plant to produce more flowers and keeps it tidy.

## deciduous

Deciduous trees, shrubs and vines begin to lose their leaves in the autumn and have none during the winter when the plant is dormant. New growth will be stimulated by the warmer and longer days of spring.

## dams

A dam is an artificial pond built to store water for irrigation, domestic use and livestock. The structure should have a sound design so that it is correctly placed, is stable and relatively impermeable.

Healthy dams, sometimes called 'enhanced dams', are usually fenced and have healthy native vegetation, including aquatic plants, in and around the dam. The result is that the water is cleaner and the new ecosystem attracts wildlife like birds and frogs. It can be much like a wetland in some instances.

DAMS

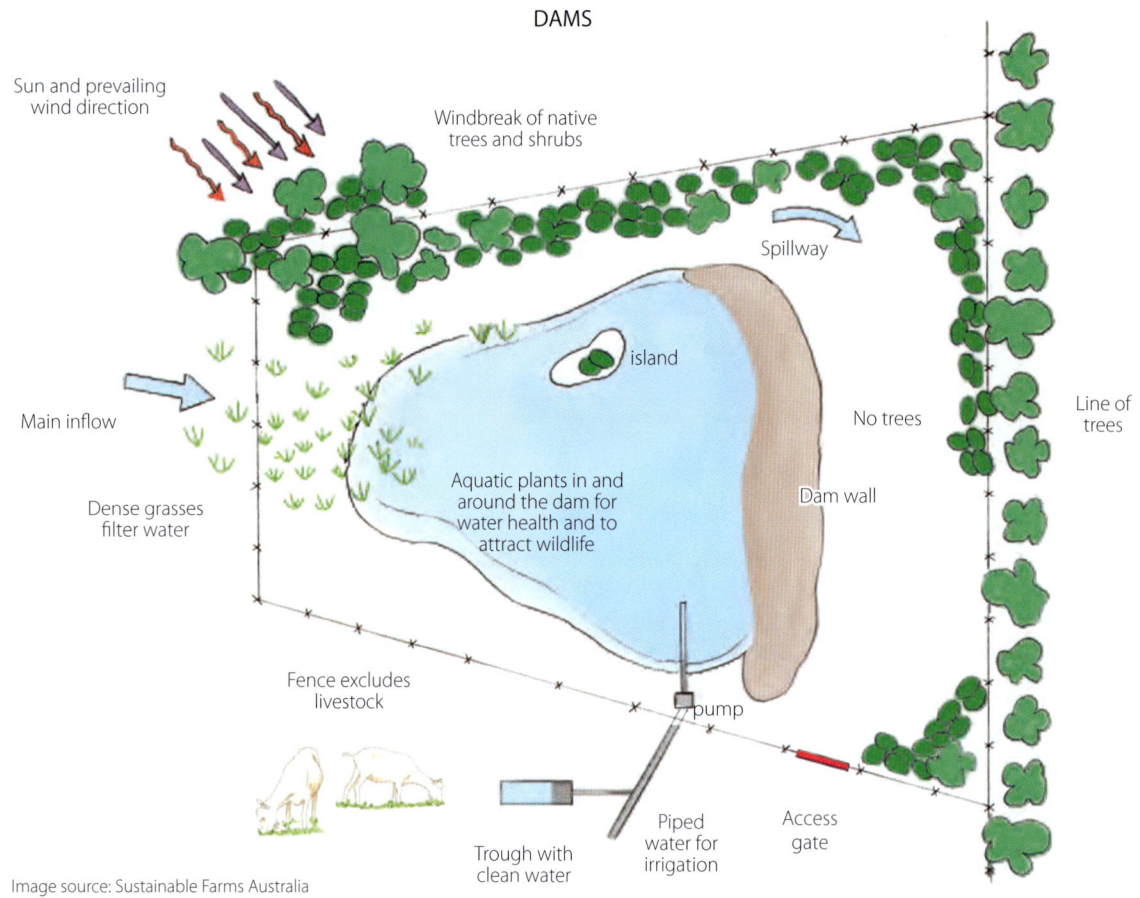

Sun and prevailing wind direction

Windbreak of native trees and shrubs

Spillway

island

Main inflow

No trees

Line of trees

Dense grasses filter water

Aquatic plants in and around the dam for water health and to attract wildlife

Dam wall

Fence excludes livestock

pump

Trough with clean water

Piped water for irrigation

Access gate

Image source: Sustainable Farms Australia

## determinate growth

Determinate growth refers to a flowering (and fruiting) shoot that will grow vertically at first, then stop. The remainder of the growth will take place on side shoots. These are also limited in their growth, giving them a 'dwarf' size or bush-like shape.

Flowers blossom at the tip of the shoots first, then form fruits that usually ripen and are harvested at about the same time.

Plants with determinate growth include tomatoes, cucumbers and bush beans.

*cf.* **indeterminate growth**

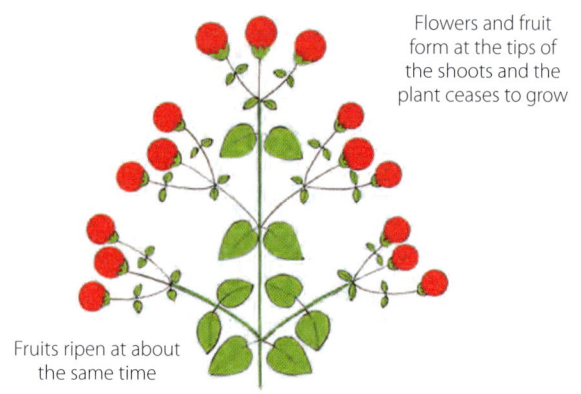

Flowers and fruit form at the tips of the shoots and the plant ceases to grow

Fruits ripen at about the same time

Some plants can have either determinate or indeterminate growth forms. These include tomatoes, cucumbers, peas and beans.

### diatomaceous earth (DE)

Diatomaceous earth is the fossilised remains of prehistoric freshwater, single-celled algae, called diatoms.

The crumbly powder is mined from ancient riverbeds and lakes and forms a soft powder that can be used as a pesticide. There are two forms, pool grade and food grade. Only food grade DE is suitable for gardens.

It works by coating an insect's body with tiny barbed cylinders that cause injuries and kills the pest. DE should be used in a targeted way. It works on 'good' and 'bad' insects and should be applied carefully. Avoid flowers where bees feed.

DE has a long-standing track record for safety and effectiveness. It is inexpensive and has little or no toxicity to the surrounding environment if used correctly.

### dibber, dibble stick, dibbler

A dibber is a stick used to make a hole in the soil for planting seedlings, cuttings, bulbs, etc.

The thickness of the stick depends on the task. A small dibble stick is also used to loosen the soil and gently lift seedlings when transplanting. They are also available commercially, and some have measurements along the shaft to indicate the depth of the hole for planting.

dibber stick

Commercial dibber with measurements

### dicotyledons

Angiosperms were formerly divided into monocotyledons, with one seed leaf (cotyledon) in the seed, and dicotyledons, mostly with two leaves in the seed.

Dicotyledons are no longer regarded as a natural grouping. Angiosperms are now divided into several groups, the larger groups being eudicots (with pollen grains having three pores or furrows) and monocotyledons (with pollen grains having one pore or furrow).

### dig and drop composting

Dig and drop composting, also called pit composting and trench composting, is done by burying kitchen waste in a trench or hole in the soil.

Pit composting is useful in established perennial gardens. Organic matter breaks down in the soil and delivers nutrients directly to the roots of nearby plants.

Trench composting can be used for larger amounts of organic material. After 6 months or so, seeds and plants can be grown along the sides of the trench.

Dig a hole or trench at least 25 cm deep and as wide as you need. Then replace the soil to cover the food scraps. Avoid burying meat and dairy products as they will encourage animals and pests into your garden.

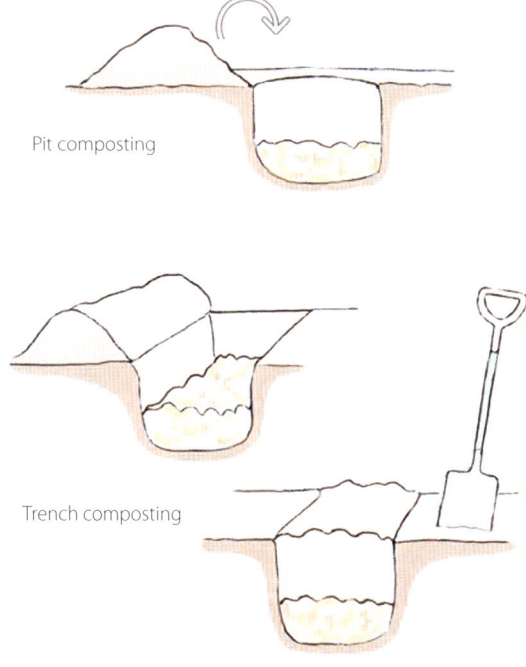

Pit composting

Trench composting

## dioecious

Having male and female flowers on different plants, like asparagus, holly and dates.

## disbudding

Disbudding is the removal of flower buds so the plant can use its energy to produce large blooms.

## division

Division is the splitting of a parent plant into two or more smaller parts for propagation. The new plants will be clones of the parent.

Clumps, stolons, rhizomes and crowns can be divided. This is done with a sharp knife, or larger plants like rhubarb are cut with an axe or sharp spade.

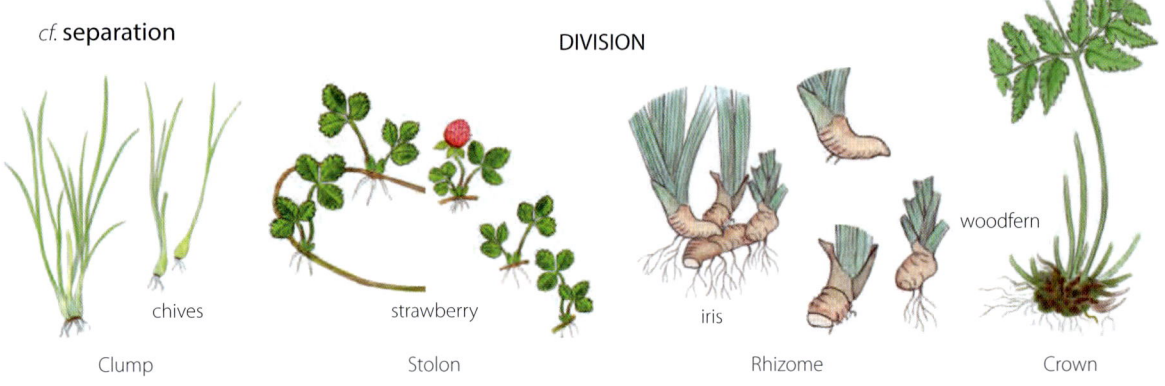

*cf.* **separation**                                        DIVISION

woodfern

chives                    strawberry                    iris

Clump                    Stolon                    Rhizome                    Crown

## dormant

A plant is dormant when its buds, seeds or other reproductive parts are alive but temporarily inactive. Herbs that die back to bulbs or rhizomes and trees that are deciduous, are said to be dormant.

Seeds are dormant prior to germination. Dormancy can be broken by the seed itself when there is sufficient moisture, warmth, oxygen, and, for some seeds, light.

Some seeds are inhibited by a particularly hard seed coat or conditions inside the seed itself that prolong dormancy. Nature helps break dormancy by scarification and stratification.

## double digging

A method used to increase and improve the depth of soil that will be cultivated when establishing a new garden bed. Double digging is done only once, as it destroys the soil structure.

To begin with, the garden bed is watered thoroughly over a few days before digging. A layer of topsoil is removed to make a trench about 30 cm wide. This exposes the subsoil or hard pan beneath. The subsoil is broken up, and a thick layer of organic matter is dug in. The topsoil from the next trench is placed on the amended subsoil. Once the number of trenches is completed, the garden is raked smooth and covered with a layer of compost.

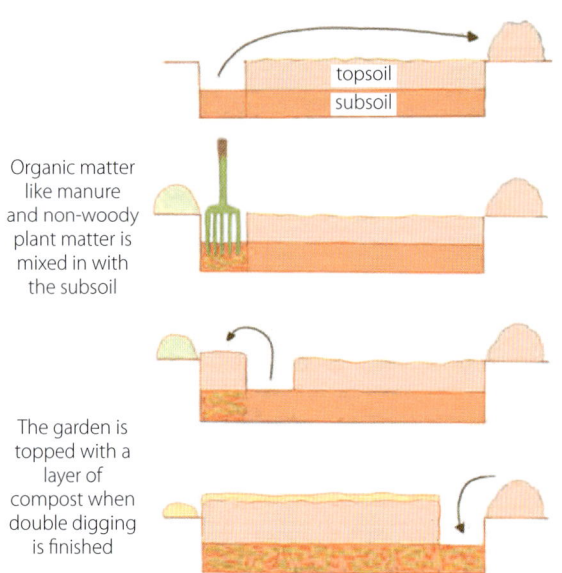

topsoil
subsoil

Organic matter like manure and non-woody plant matter is mixed in with the subsoil

The garden is topped with a layer of compost when double digging is finished

## double flower

A flower with more than the usual number of petals, as some peonies (*Paeonia*), or with petal-like infertile stamens (staminodes) in addition to normal petals, as the canna lily (*Canna indica*).

A single flower has one row of petals around the centre of the flower. Genetic mutations in a single flower produce double flowers with multiple layers of petals that hide the centre of the flower, and semi-double flowers that have additional petals with the centre of the flower still visible.

Many of the most beautiful flowers are double-flowered cultivars that have been bred for their beauty. These include hybrid tea roses, camellias and carnations. Double flowers are usually infertile. Bees find it difficult to pollinate them, and some have stamens and styles that have reverted to petals.

## double working

In grafting, double working uses two graft unions instead of one. It is done when the scion and rootstock are incompatible. An interstock that is compatible with both is inserted between the scion and the rootstock.

There are two stages. First, the interstock is grafted onto the rootstock. When that graft is established, the scion is grafted onto the interstock.

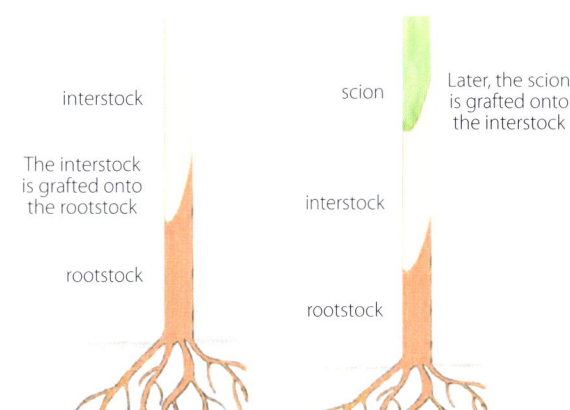

interstock

The interstock is grafted onto the rootstock

rootstock

scion

Later, the scion is grafted onto the interstock

interstock

rootstock

## drip irrigation *see* irrigation

## drip line

A drip line is an imaginary circle on the ground representing the perimeter of the tree canopy.

The drip line is also called the Critical Root Zone. Most water is absorbed here, not near the trunk. Tiny tree rootlets on the drip line take up water and nutrients from the soil. The drip line is the best place to water and apply fertiliser. Mulch can cover the drip line and extend under the tree up to 30–50 cm from the tree trunk. Disturbing the drip line, for example, with trenching or paving, is likely to harm the tree.

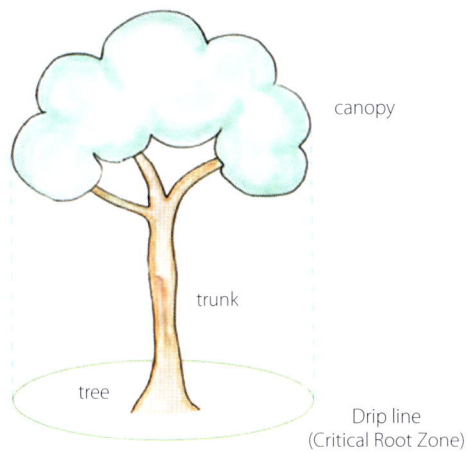

canopy

trunk

tree

Drip line
(Critical Root Zone)

## drupe

A drupe is a fruit that usually contains a single seed inside a stone.

'Stone fruit' is a general term for *Prunus* species that have fleshy drupes, like peaches, apricots, cherries, plums and prunes. Almonds and coconuts have drupes with a tough rather than a fleshy layer. Mangos, walnuts and dogwoods also bear drupes.

Raspberry and blackberry fruits are clusters of tiny drupes (drupelets) and are not berries at all.

almond
(*Prunus dulcis*)

Pitted shell encloses the edible seed

apricot
(*Prunus armeniaca*)

Stone encloses the seed. Fleshy layer around the stone is edible.

raspberry
(*Rubus idaeus*)

Fruit is a cluster of drupelets

### earthworms

There are three kinds of earthworm.

Composting worms move around randomly and feed in the organically rich surface layer of the soil. These worms are used in worm farms.

The second group of worms make horizontal burrows and live within 5–30 cm below the soil surface, feeding on organic matter that they ingest with soil. They are sometimes found in open-bottomed compost bins.

The third group of worms is rarely seen. They construct deep vertical burrows (to 2 m) in the soil profile. Their holes open onto the soil surface where they come to feed at night. They move organic matter and nutrients down into the soil, creating channels for air and water and leaving castings around roots as they move around.

Soil-living earthworms till the soil and aerate it with their burrows. Their burrows aid root and water penetration. All worms excrete fertile castings with readily available nutrients for plants.

Some 6000 earthworm species have been identified. They are native to a particular area. Species from elsewhere that are invasive can cause environmental problems.

*see also* **worm farming**

**EARTHWORMS**

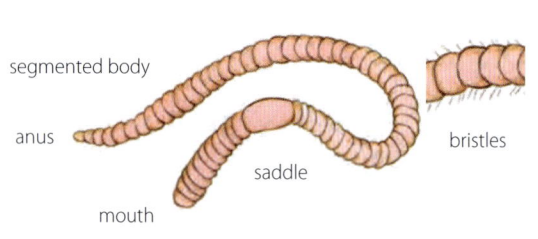

segmented body

anus

mouth

saddle

bristles

All earthworms have common characteristics

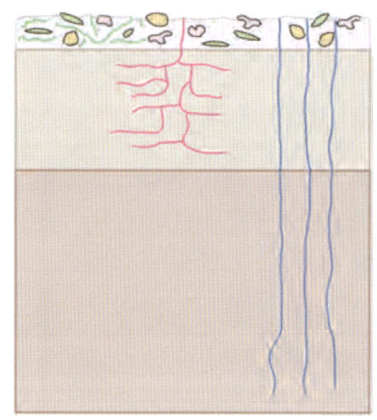

Composting worms mostly in surface litter

Soil-living worms with horizontal burrows

Soil-living worms with deep vertical burrows

### ecological gardening

An ecological garden has a low impact on the environment, is low maintenance once established and has high yields of produce. It is grounded in working with nature and includes permaculture, organic gardening and sometimes, indigenous gardening practices.

In nature the soil is not tilled and there is a mix of plants in a given area rather than a single crop. Over time, the undisturbed soil holds water and becomes rich with a thriving web of life that builds nutrients. Above ground there is a flourishing ecosystem of edible plants that can feed people, pollinating insects and birds. Some gardens eventually become self-sustaining. Gardeners make use of mulch and compost and incorporate other systems like succession planting, companion planting and guilds. Crop rotation is not necessary when there is a dense planting of different species.

*see also* **food forest gardens, guilds, permaculture**

### egg cell

The egg cell is the female reproductive cell of a seed plant. It unites with a male sperm cell at fertilisation.

The egg cell is located in an ovule. Once fertilisation has occurred, the ovule will give rise to a seed. In flowering plants (angiosperms) it is enclosed in an ovary. In non-flowering seed plants (gymnosperms) it is usually exposed on a cone scale.

## elevated garden beds

Elevated garden beds are rectangular boxes on legs, filled with soil. They are no wider than 1.2 m, so that it is easy to reach inside the bed.

They are used in urban areas where there is no soil and are useful for people who don't want to bend down or are disabled.

Elevated garden bed on legs

## epigeal germination *see* germination

## epiphyte

An epiphyte is a plant that can provide its own nutrients but grows on another plant for support.

Many tropical orchids, bromeliads and ferns are epiphytes.

Epiphytic staghorn fern (*Platycerium bifurcatum*)

## erosion *see* soil erosion

## espalier

An espalier is a fruit tree or ornamental shrub with branches trained to grow flat against a wall or fence, or on a lattice.

Several different plants can be espaliered in a small space. There are many espalier patterns, some more natural and others more formal.

An espalier is pruned each year to train and maintain its shape.

## etiolation

Etiolation, also commonly known as 'legginess', is the excessive lengthening of the shoots of a plant as it reaches for a source of light.

An etiolated seedling will be spindly, with a weak pale stem.

## eudicots

Eudicots are the largest group of flowering plants (75% of all angiosperms).

They are characterised by a seed with two cotyledons (seed leaves); flower parts often in multiples of four or five, and leaves with lattice-like veins. Typically, the root is a taproot. Plants are herbaceous or woody.

*cf.* dicotyledons, monocotyledons

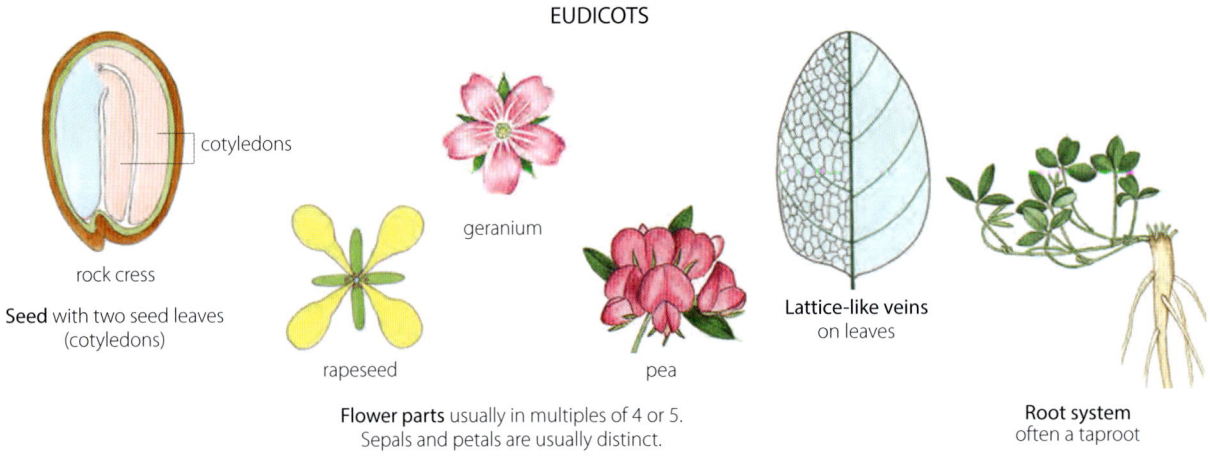

**EUDICOTS**

cotyledons

rock cress

**Seed** with two seed leaves
(cotyledons)

geranium

rapeseed

pea

**Flower parts** usually in multiples of 4 or 5.
Sepals and petals are usually distinct.

Lattice-like veins
on leaves

**Root system**
often a taproot

---

## evergreen

Evergreen trees, shrubs and vines do not lose all of their leaves in winter. Some leaves are shed throughout the year as a result of age, stress and/or disease.

## exotic plants

Exotic plants are species that are not native to a particular place.

They may be introduced into an area from a different part of the country or from another country altogether. Popular introduced species in Australia include kiwifruit (*Actinidia deliciosa*) that is native to mainland China and Taiwan, and tulip species (*Tulipa*) that are native to Central Asia and Turkey.

Some exotic species can become weedy. The Cootamundra wattle (*Acacia baileyana*) is native to the Temora-Cootamundra district in New South Wales, but has become an environmental weed in other areas of Australia.

## eye

A small depression on a stem tuber, like a potato.

Each eye represents a node on a stem and bears a scale leaf and one or more buds from which a new plant can grow.

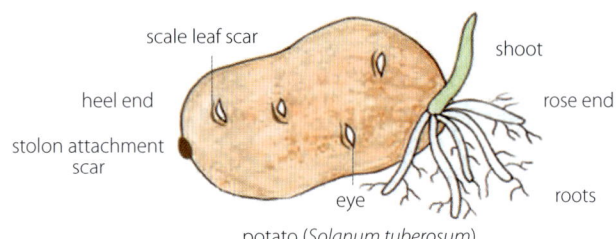

scale leaf scar

heel end

stolon attachment
scar

shoot

rose end

eye

roots

potato (*Solanum tuberosum*)

## F1 hybrid

An F1 hybrid results from two pure parent plant lines, with constant features as a result of inbreeding, that are cross-pollinated under controlled conditions. The female parent, or seed parent, bears the F1 hybrid seeds that will combine the desired features of both parents.

Cross-fertilisation is usually done by hand, and the whole process is an expensive one. Seeds of F1 hybrid plants do not grow true to type and new seeds must be bought and sown each year.

The parent lines lose vigour during the inbreeding process, but this is restored in the F1 hybrid plant and is known as 'hybrid vigour'.

# family

A family is a group of one or more genera that share common features. The names of families end in 'aceae'. Apples (*Malus*), almonds and cherries (*Prunus*), roses (*Rosa*) and raspberries (*Rubus*) are all members of the Rosaceae family. Plants in Rosaceae usually have symmetrical, bisexual flowers with five petals, and many stamens.

*see also* **plant kingdom**

## SOME COMMON GARDEN FAMILIES AND GENERA

**AMARYLLIDACEAE**
amaryllus family

| | |
|---|---|
| *Allium* | onion, garlic |
| *Crinum* | crinum |
| *Narcissus* | daffodil |

onion · daffodil

**APIACEAE**
carrot family

| | |
|---|---|
| *Apium* | celery |
| *Daucus* | carrot |
| *Petroselinum* | parsley |

carrot

**ASTERACEAE**
daisy family

| | |
|---|---|
| *Matricaria* | chamomile |
| *Helianthus* | sunflower |
| *Lactuca* | lettuce |

chamomile

**BRASSICACEAE**
mustard family

| | |
|---|---|
| *Brassica* | cauliflower, broccoli |
| *Tropaeolum* | nasturtium |

nasturtium · broccoli

**CACTACEAE**
cactus family

| | |
|---|---|
| *Mammilaria* | pincushion cactus |
| *Opuntia* | prickly pear |

prickly pear · pincushion cactus

**CONVOLVULACEAE**
bindweed family

| | |
|---|---|
| *Convolvulus* | bindweed |
| *Ipomoea* | sweet potato |

sweet potato

**CUCURBITACEAE**
gourd family

| | |
|---|---|
| *Cucumis* | cucumber |
| *Cucurbita* | pumpkin |
| *Lagenaria* | bottle gourd |

pumpkin

**FABACEAE**
pea family

| | |
|---|---|
| *Acacia* | wattle |
| *Pisum* | pea |
| *Medicago* | alfalfa |

pea · wattle

**IRIDACEAE**
iris family

| | |
|---|---|
| *Crocus* | crocus |
| *Gladiolus* | gladiolus |
| *Iris* | iris |

iris · crocus

**LAMIACEAE**
mint family

| | |
|---|---|
| *Lavandula* | lavender |
| *Mentha* | mint |
| *Salvia* | sage |
| *Thymus* | thyme |

mint · thyme

**LILIACEAE**
lily family

| | |
|---|---|
| *Fritillaria* | fritillary |
| *Lilium* | lily |
| *Tulipa* | tulip |

lily · tulip

**ORCHIDACEAE**
orchid family

| | |
|---|---|
| *Dendrobium* | dendrobium |
| *Paphiopedilum* | slipper orchid |

slipper orchid

**POACEAE**
grass family

| | |
|---|---|
| *Arundinaria* | bamboo |
| *Saccharum* | sugarcane |
| *Triticum* | wheat |
| *Zea* | corn |

corn · bamboo

**RUTACEAE**
citrus family

| | |
|---|---|
| *Citrus* | lemon, lime |
| *Boronia* | boronia |

lemon

**ROSACEAE**
rose family

| | |
|---|---|
| *Malus* | apple |
| *Prunus* | almond, cherry |
| *Rosa* | rose |
| *Rubus* | raspberry |

raspberry · rose · almond

**SOLANACEAE**
nightshade family

| | |
|---|---|
| *Solanum* | potato, tomato |
| *Capsicum* | peppers |

tomato · pepper

## fedge

A fedge is a self-sustaining food-producing hedgerow. The term also refers to a hedge that is used for a fence.

Plants are perennial and pruned to maintain an accessible size. Prunings are chopped and dropped for mulch. A subtropical fedge might include banana, avocado, macadamia, mulberry, lychee and pigeon pea.

## ferns

Ferns are an ancient group of plants with fossil records showing they have been around for some 300 million years. Today there are over 10 000 known fern species. They are green flowerless plants that reproduce by spores rather than seeds. Their leaves (fronds) usually arise from an underground rhizome, above-ground stolon or an erect trunk (tree ferns). Ferns grow from the tropics to the Arctic but most prefer indirect sunlight, moist soil and high humidity.

There are many ways to propagate ferns. Spores, the equivalent of seeds in flowering plants, are borne in sporangia in sori on the underside of a frond. It can take some months for spores to develop into a plant. However, the maidenhair fern (*Adiantum*) can take as little as 6 weeks. The slow-growing and impressive austral king fern (*Todea barbara*), native to Australia, New Zealand and South Africa, is propagated from spores by specialist horticulturists.

Ferns can be reproduced vegetatively. The hen and chicken fern (*Asplenium bulbiferum*) develops tiny plantlets called bulbils on the fronds that will take root easily on moist soil. Other ferns are propagated by division of the creeping rhizome or stolon. The mat-forming sword fern (*Nephrolepis cordifolia*) creeps by stolons and has dense clumps that can be separated and grown. The slow-growing and impressive austral king fern (*Todea barbara*), native to Australia, New Zealand and South Africa, sometimes forms a trunk with several crowns. It is propagated from spores by specialist horticulturists. Tree ferns (*Cyathea*) can be propagated by offshoots that emerge from the base of the parent plant.

The staghorn fern grows on other plants. Its shield-shaped infertile fronds lie against the tree and protect the small rhizome and its roots. They make colonies when their rhizomes spread. Spores from antler-shaped fertile fronds are windblown and germinate naturally on surrounding trees.

The climbing fern (*Lygodium*) is sold as an ornamental plant. The fast-growing Japanese climbing fern (*Lygodium japonicum*) has escaped cultivation in gardens and has become an environmental weed in New South Wales and south-eastern Queensland.

**FERNS**

hen and chicken fern
(*Asplenium*)

sword fern
(*Nephrolepis*)

maidenhair fern
(*Adiantum*)

tree fern
(*Cyathea*)

bracken
(*Pteridium*)

king fern
(*Todea*)

staghorn fern
(*Platycerium*)

climbing fern
(*Lygodium*)

sorus with sporangia

sporangium

spores

frond

back of frond with sori

crozier or fiddlehead

rhizome with roots

**Fern**

## fertilisation

In seed plants, angiosperms and gymnosperms, the fusion of a male sperm cell in pollen, with a female egg cell. The resulting cell will become a seed.

In flowering plants (angiosperms) cross-fertilisation occurs between flowers from separate plants, and self-fertilisation occurs between flowers on the same plant or within the same flower.

*cf.* pollination

## fertiliser

Fertilisers increase the availability of nutrients for plant growth.

They are spread on top of soil or mulch and watered in. Liquid fertilisers that are sprayed on leaves are fast-acting, but are only a temporary measure.

Natural fertilisers that are organic are made from plant or animal waste and include manures, composts, and animal by-products like blood and bone, as well as some that are inorganic like potassium and lime. They are broken down by microorganisms, and the nutrients are released more slowly than those in synthetic fertilisers. The steady release over time makes them available for longer than synthetic fertilisers.

Synthetic fertilisers are made through industrial processes or mined from deposits in the earth. They are purified and blended for easy application and are rapidly available to the plant.

Fertilisers are usually labelled with three numbers that give the percentage by weight of nitrogen (N), phosphorus (P) and potassium (K).

Natural fertilisers improve the soil that feeds the plant, while synthetic fertilisers added to the soil directly feed the plant.

## festooning

Festooning is a method of bending down the branches of fruit trees to induce fruiting rather than new growth.

The young branches are weighted or tied down until they remain in position naturally. Fruit is closer to the ground and easier to harvest. Dwarf fruit trees can also be festooned.

## filament *see* flower

## flats

A flat is a shallow rectangular tray for growing seedlings or for holding seed trays with cells.

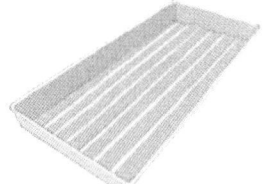

## floricane *see* bramble

## flower

The flower is the structure for sexual reproduction in flowering plants (angiosperms).

It is typically arranged in whorls. The outer protective whorl of sepals is the calyx. Next is the whorl of petals, the corolla, that surrounds the reproductive structures. A whorl of male stamens typically surrounds the female pistil or pistils.

The stamen consists of a filament that supports the pollen-bearing anther. The female pistil typically has a stigma that receives pollen at the tip. The stigma is usually connected to the ovary by a style. The whorls are all attached to a receptacle at the tip of the flower stem.

The pollen-bearing anther contains the male sex cells (pollen). The ovary of the pistil contains the female sex cells in the ovules.

**FLOWER**

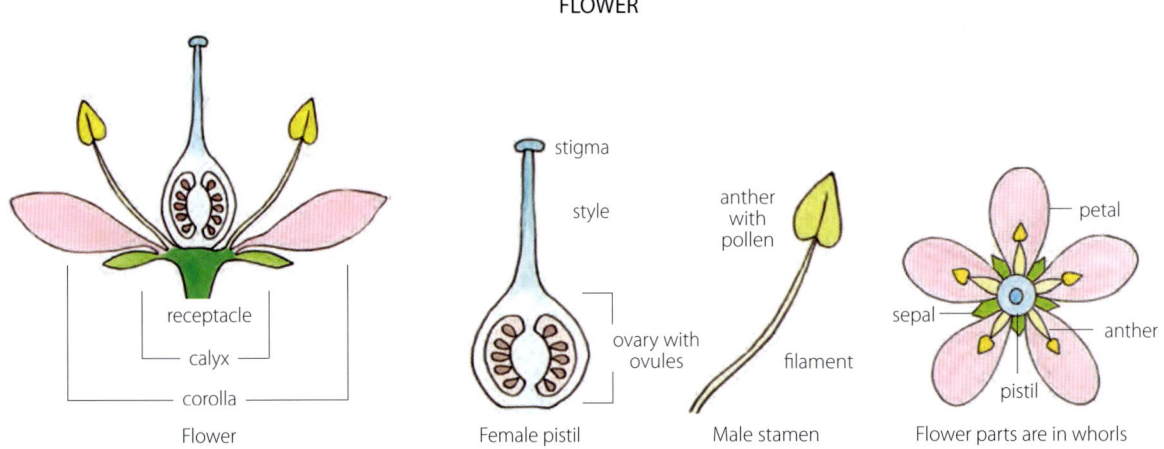

Flower

Female pistil

Male stamen

Flower parts are in whorls

### foliar feeding

Plants can absorb nutrients through their foliage. A diluted liquid feed, such as seaweed, worm tea, comfrey tea or a commercial product, is sprayed on leaves on a still, rainless day.

Foliar feeds are usually applied in the morning when the leaf pores (stomata) are open. The timing of application may be important, such as at or after flowering, or during fruiting. Foliar feeding is fast-acting and delivers nutrients through the leaves immediately.

strawberry

raspberry

Comfrey tea as a foliar spray is rich in nitrogen, phosphorus and potassium. It benefits many plants, including tomatoes, eggplants and berries.

### food forest gardens *see* page 53

### forcing

Forcing is a technique used to make plants flower or produce fruit outside their normal growing season. This requires an artificial environment, like a greenhouse. Hotbeds and greenhouses are used to raise seedlings early so they will flower and fruit earlier than usual.

Some spring-flowering bulbs can be forced to flower early by placing the bulb in water or by chilling the bulb before growing.

### friable soil

Friable soil has a crumbly texture and enough air space to allow drainage but also trap nutrients. This texture is ideal for garden soil.

### frond *see* ferns, palms

# food forest gardens

A forest in nature can have up to seven layers: canopy trees, smaller understorey trees, shrubs, herbaceous plants, ground cover, roots, and plants that grow vertically. A food forest garden imitates the layers, the difference being that most of the plants are edible.

Food forest gardens are managed by people and are largely self-sustaining. They date back hundreds, even thousands, of years. Desert oases are food forests. The oasis at Inraren in Morocco is thought to be 2000 years old. The floating gardens (chinampas) in Mexico were built by the Aztecs and flourished for about 100 years, from the early 15th century until the Spanish conquest in the 16th century.

Food forests inspired the permaculture movement that started in the 1970s and is now worldwide.

Seven layers of a food forest

1. **Canopy**    Large nut and fruit trees like date palms, walnuts and mulberries

2. **Understorey**    Small trees like olives, peaches, apples and citrus

3. **Shrubs**    Low, branched, woody plants like rosemary, lavender, blueberry and raspberry

4. **Herbaceous plants**    Non-woody plants like vegetables and herbs

5. **Climbers and vines**    Kiwifruit, grapes and climbing beans

6. **Ground cover**    Prostrate thyme, rosemary and strawberries

7. **Roots**    Root vegetables, like carrots and parsnips, and spices, like ginger and turmeric

FOOD FOREST GARDENS

Some plants of a Moroccan oasis

1. Canopy
5. Climbers
2. Understorey
3. Shrubs
4. Herbs
6. Ground cover
7. Roots

Forest layers

1. Canopy
date palm

carob

pomegranate

2. Understorey

desert mint

onions

alfalfa

4. Herbaceous layer

olive

argan nut tree

2. Understorey

orange

melons

lavender

3. Shrubs

grapes

turmeric

Animals

prostrate thyme

6. Ground cover

5. Climbers and vines

7. Roots

Local sheep breeds, like the d'man ewe that lives in an enclosure and feeds on alfalfa. It provides meat and can lamb twice a year.

## frost

Frost forms in two ways: either by frozen water vapour in the air being deposited on plants, or by dew droplets on plants freezing to form ice crystals. It occurs when overnight temperatures fall below 0°C (32°F), the weather is calm, usually the skies are cloudless, and the ground is moist.

Deciduous plants, like fruit trees, maples, magnolias and flowering cherries, avoid damage from frost by becoming dormant in winter.

Plants that are frost-tolerant include spring flowering bulbs, and some grevilleas and callistemons.

Tender plants that are damaged by frost can be protected by cloches or horticultural fleece over a garden bed or tunnel. Others will survive in a hothouse or a hotbed.

### FROST TOLERANCE OF VEGETABLES

| Frost-hardy vegetables withstand frost below −2°C (28°F) for short periods include | Frost-tolerant vegetables withstand light frost, −2°C–0°C (28°F–32°F) include | Frost-tender vegetables damaged by light frost, −2°C–0°C (28°F–32°F) include |
|---|---|---|
| broccoli | beet | basil |
| Brussels sprouts | bok choy | beans |
| cabbage | carrot | cucumber |
| kale | cauliflower | edamame |
| leeks | celery | eggplant |
| onion | chard | melons |
| parsley | Chinese cabbage | okra |
| peas | endive | pepper |
| radish | Jerusalem artichoke | pumpkin |
| rocket | lettuce (some varieties) | rosemary |
| spinach | garlic and chives | squash |
| thyme | parsnips | sweet corn |
| turnips | potatoes | sweet potato |
|  | rhubarb | tomatoes |

**frost hardy** *see* frost

**frost tender** *see* frost

**frost tolerant** *see* frost

**fruit** *see* page 55

## fruit fly

Fruit fly is a major pest worldwide. It lays its eggs in ripening fruit. The eggs hatch as larvae (maggots) that feed on the flesh, causing it to rot. The fruit falls to the ground, where the larvae burrow into the soil to pupate. Adult flies can emerge in as little as seven days.

There are two main types of fruit fly in Australia. The introduced Mediterranean fruit fly (*Ceratitis capitata*) is found in Western Australia, and the Queensland fruit fly (*Bactrocera tryoni*) is found in the other states except for Tasmania and South Australia.

Fruit fly numbers tend to increase in spring when there is a continuous availability of host plants, like stone fruit, apples, berries, tomatoes and mangoes.

## fungicide

A fungicide is a substance that kills mould and fungus.

*see also* Bordeaux mixture, pesticides

## fruit

A fruit is the mature ovary of a flower. It encloses the seeds that develop after the flower has been pollinated.

All flowering plants bear fruit and many are edible. They include apples, peas, grains, citrus, pumpkins, stone fruit and berries.

FRUIT

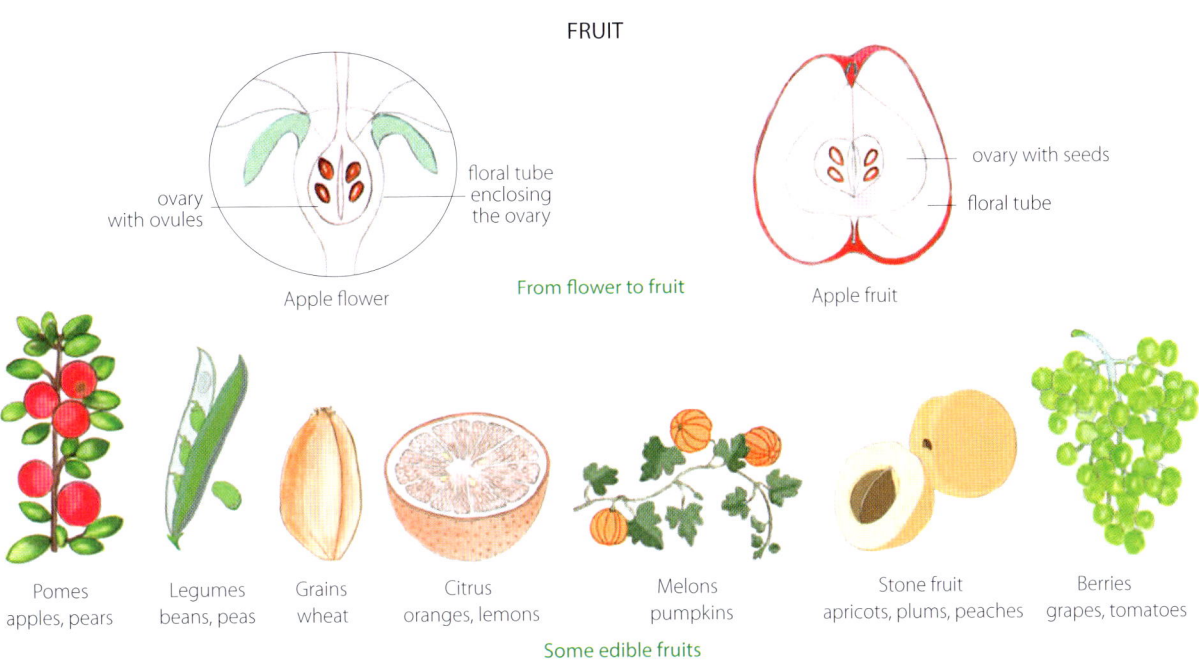

ovary with ovules

floral tube enclosing the ovary

Apple flower

From flower to fruit

ovary with seeds

floral tube

Apple fruit

Pomes
apples, pears

Legumes
beans, peas

Grains
wheat

Citrus
oranges, lemons

Melons
pumpkins

Stone fruit
apricots, plums, peaches

Berries
grapes, tomatoes

Some edible fruits

## fungus, *pl.* fungi

Fungi include moulds, mildews, yeast, mushrooms and toadstools. They are made up of thread-like hyphae that invade a food source and digest it externally before absorbing the nutrients. Together with bacteria, fungi are a primary decomposer of organic matter in the soil.

The hyphae form a tangled network called the mycelium, that covers vast areas of soil. The mycelium transports nutrients to plant roots.

The most visible part of a fungus is the fruiting body, like that of a mushroom. It is made up of hyphae that produce reproductive spores.

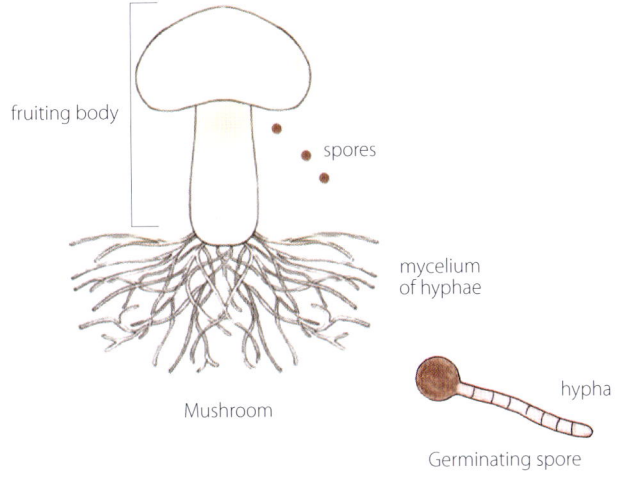

fruiting body

spores

mycelium of hyphae

Mushroom

hypha

Germinating spore

## gall

A gall is an abnormal swelling or outgrowth of plant tissue that forms due to irritation by an organism such as a wasp, midge, bacterium or virus, or by a wound.

The most well-known gall wasp in Australia is the citrus gall wasp (*Bruchophagus fellis*) that forms stem galls on all varieties of citrus. The wasp lays its eggs inside a soft new shoot, and the gall forms to make a home and food for the developing lava.

The midge that lays its eggs in beaded glasswort makes a flower-like gall

The midge lava develops inside the gall

## garden history

'The earliest documented gardens were in the area known as the Fertile Crescent, which extended from Egypt north-east through the eastern Mediterranean and then south-east to the Persian Gulf. The Garden of Eden is traditionally located in southern Mesopotamia, the eastern end of the crescent, near the confluence of the Tigris and Euphrates. ... in antiquity the rivers ran separately into the Persian Gulf through a place called 'Edin', which has the sense of fertile, arable land.' Gordon Campbell, 'A Short History of Gardens' (2016).

The first gardens were usually enclosed. 'If the original meaning of the English word was rooted in the presence of a 'gard', an edge or fence, the content of the fenced space had always been a mixture of productive and non-productive, useful and beautiful, pragmatic and symbolic. ... peasant houses (were) home to turnips and pig shit, but also to roses, healing herbs ...' Pier Vittorio Aureli and Maria Guidici, 'Gardener's World: A short history of domestication and nurturance' (2021).

Temples in the ancient world had gardens that became public spaces, as did those in ancient China and ancient Greece. Monastic gardens and secular feudal gardens flourished between the 9th and 15th centuries in Europe. Botanic gardens emerged between the 16th and 17th centuries, a time of expansion of European empires, to systematically study and name plants.

Chinese contemplative gardens were laid out to reflect the grandeur of nature, and are opposite in style to the symmetrical Baroque gardens in Europe. Fine gardens in history were a mark of the status of princes and monarchs. The permaculture movement that began in the 1970s, however, was inspired by ancient food forests, the aim being to make long-term, self-sustaining gardens anywhere in the world. Today, gardens like these, and domestic and community gardens in many forms, are promoted everywhere as a necessary way of feeding the world's population.

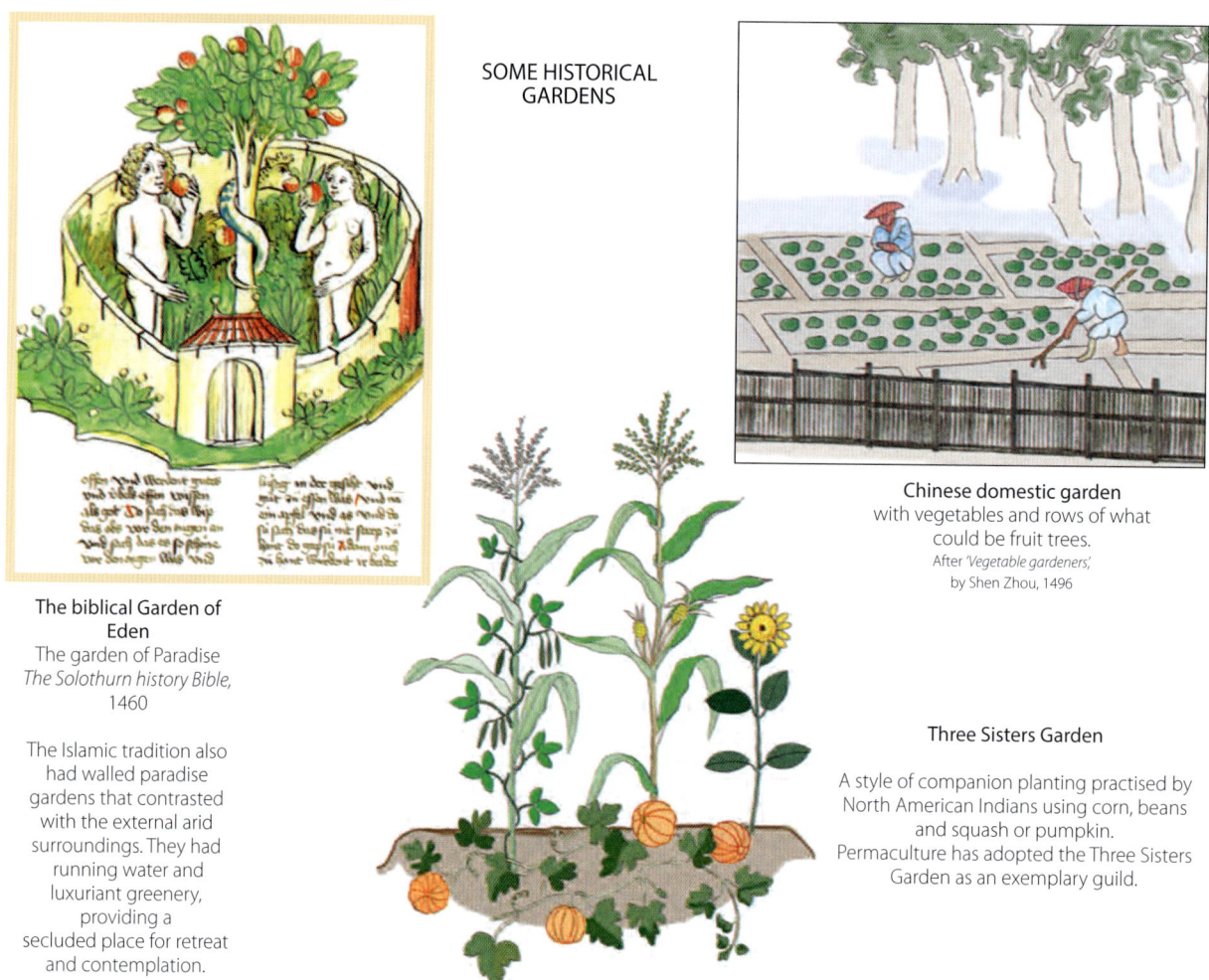

SOME HISTORICAL GARDENS

**The biblical Garden of Eden**
The garden of Paradise
*The Solothurn history Bible,*
1460

The Islamic tradition also had walled paradise gardens that contrasted with the external arid surroundings. They had running water and luxuriant greenery, providing a secluded place for retreat and contemplation.

**Chinese domestic garden**
with vegetables and rows of what could be fruit trees.
After *'Vegetable gardeners,'*
by Shen Zhou, 1496

**Three Sisters Garden**

A style of companion planting practised by North American Indians using corn, beans and squash or pumpkin.
Permaculture has adopted the Three Sisters Garden as an exemplary guild.

**genetically modified seeds** *see* mutation

**genus,** *pl.* **genera**
In plants, is a group of one or more species that share similar or distinct characteristics. Related genera are more broadly grouped into a family.

Almonds and cherries belong in the genus *Prunus* and raspberries and blackberries in the genus *Rubus*. These genera have characteristics in common and are in the rose family Rosaceae. The apple *Malus domestica*, has the genus name (*Malus*) and the species name (*domestica*). It too is a member of the rose family Rosaceae.

*see also* **family, plant kingdom, species**

**germination**
Germination is the series of events that transforms a viable seed from its dormant state into a self-sustaining, growing seedling.

There are two main types of germination in plants: above ground germination (epigeal) and in ground germination (hypogeal).

The first step in germination is the rapid absorption of water. The water activates enzymes in the seed that initiate growth using food stored in the cotyledons of the seed. The seed coat ruptures, the primary root (radical) emerges, and the plumule develops into a shoot. This is the end of germination. True leaves then develop. The plant now makes its own food by photosynthesis and grows into a seedling.

All seeds have a temperature range that is required for successful germination. The soil should be moist and is usually not watered until the seedling emerges. Over-watering can compact the soil. Seeds respire like any other living organism and need an open, aerated soil for a supply of oxygen and to allow carbon dioxide to escape.

### GERMINATION OF EUDICOTS

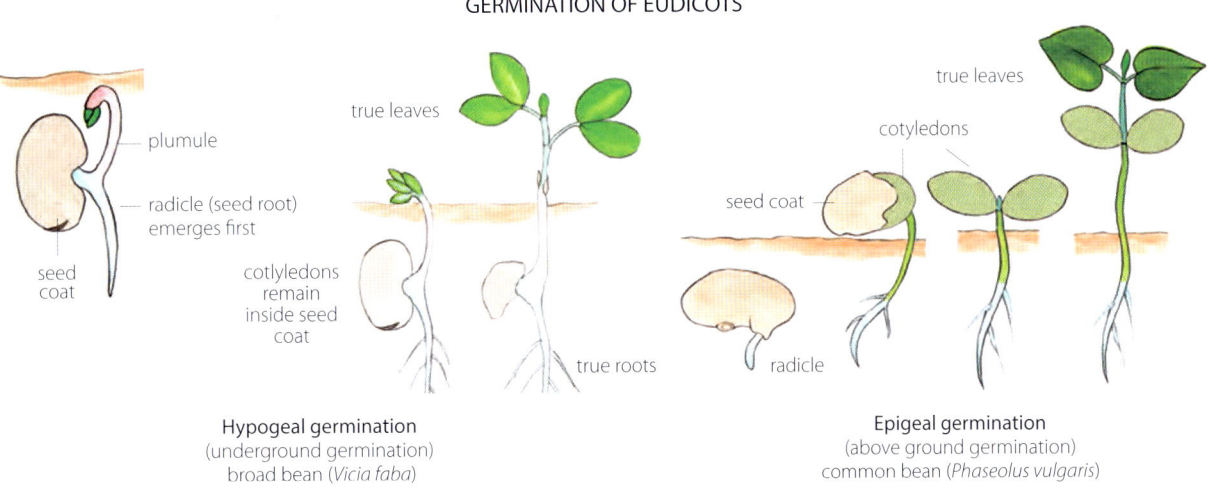

Hypogeal germination
(underground germination)
broad bean (*Vicia faba*)

Epigeal germination
(above ground germination)
common bean (*Phaseolus vulgaris*)

**girdling**
Girdling is the removal of a strip of bark from around the trunk or branch of a tree.

During the process, the phloem and cambium layers are removed, blocking the flow of nutrients from the foliage downward. The xylem, however, is left intact. Water and nutrients continue to flow upward through the xylem tissue.

*see also* **air layering**

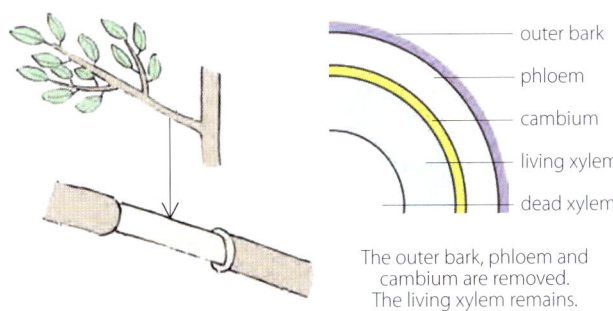

The outer bark, phloem and cambium are removed. The living xylem remains.

57

**glass house** *see* greenhouse

## graft

The term 'graft' has different meanings in horticulture.

The point of union of the scion with the rootstock is called a graft. A graft is also a bud, or a shoot with two or more buds, to be implanted into a rootstock. The term graft is used for a new plant produced from the union of two plant parts (the scion and the rootstock).

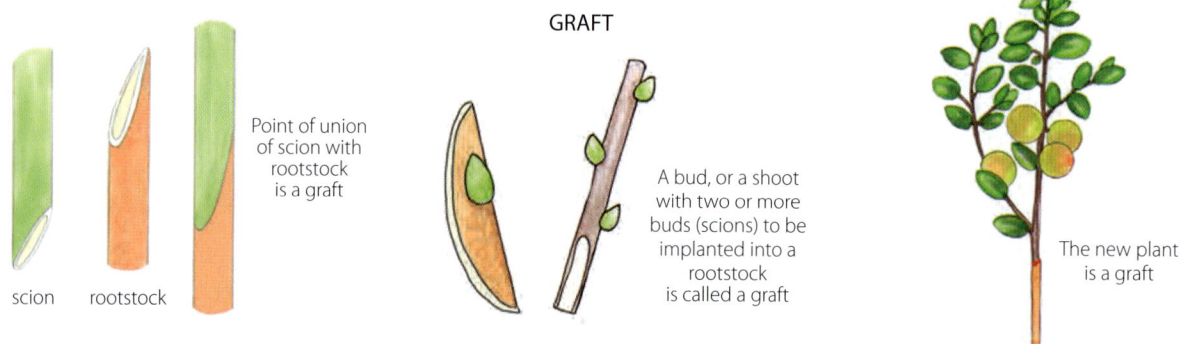

GRAFT

scion   rootstock

Point of union of scion with rootstock is a graft

A bud, or a shoot with two or more buds (scions) to be implanted into a rootstock is called a graft

The new plant is a graft

## graft chimaera, graft chimera

In grafting, a graft chimaera is a bud that appears at the junction of the scion and the rootstock. It contains tissues from both the rootstock and the scion.

Adam's laburnum (+ *Laburnocytisus* 'Adamii') is a graft chimaera between laburnum (*Laburnum*) and broom (*Cytisus*). Graft chimaeras are propagated by cloning. Graft chimaeras, sometimes called graft hybrids, are not true hybrids.

## graft compatibility

Grafts are more likely to be successful when the plants are closely related botanically.

Grafts between the same species are likely, but not always, to be successful. Nearly all *Citrus* species are compatible. However, stone fruit species in the genus *Prunus* are not all compatible. Almond (*Prunus dulcis*) grafted onto peach (*P. persica*) is compatible, whereas almond onto apricot (*P. armeniaca*) is not.

Grafts between two genera from the same family are less likely to be successful. However, quince (genus *Cydonia*) may be used as a dwarfing rootstock for pears (genus *Pyrus*).

## graft hybrid = graft chimera

## graft union

The graft union is where the tissue of the rootstock becomes continuous with the tissue of the scion. It appears as a raised scar on the stem.

When grafting, the cambium of the scion and rootstock must be in contact. The callus bridge that develops at the area of contact makes new sapwood cells and new inner bark cells. These unite the scion and rootstock so that minerals and water can flow upward through the sapwood (xylem) and sugars from photosynthesis can flow downward through the inner bark (phloem) of the new graft.

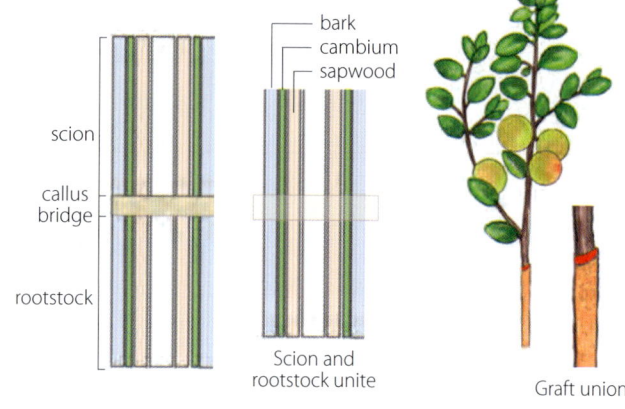

scion

callus bridge

rootstock

bark
cambium
sapwood

Scion and rootstock unite

Graft union

## graftage

Graftage is the uniting of two plant parts to make a new plant.

A bud or shoot (scion) of a species, variety or cultivar is united with a plant that is compatible and has an established root system (rootstock).

Graftage includes both budding and grafting and is mostly done on woody plants.

Green-grafting of some herbaceous plants, like tomatoes, to improve disease resistance and vigour has become widespread.

## grafting

Grafting is a method of joining parts from two different plants so that they will unite and grow as one plant.

One part is the rootstock that will become the root system of the new plant. The other is the scion, a piece of stem with buds, that becomes the above-ground growth and yields fruit.

Grafting is done on woody plants but also on herbaceous plants, like tomatoes and melons.

There are many grafting techniques. For grafting to be successful, the scion and rootstock must be compatible, they must be at the right stage of growth, and the cambium of the scion and rootstock must meet.

Grafting has many uses. It is used for plants that do not grow true from seed, such as cultivars and many fruit trees, or plants that do not root easily from cuttings.

Specific rootstocks can be chosen to suit different soils. Grafting can also be used to control disease and the size of trees. Less vigorous rootstock is used for dwarf trees than for standard trees. It can also save space by allowing two or three grafts on one rootstock. Grafts can also be selected to improve the quantity and quality of fruiting and to promote early flowering.

A tree that is no longer fruiting well can be reinvigorated by cutting it back and grafting the rootstock with a new scion. In some forms, like inarching and bridge grafting, it can be used to repair an existing tree.

Grafting is also used as a general term in the same sense as graftage.

*see also* **budding, graftage, green-grafting**

### GRAFTING TECHNIQUES

| rootstock | scion | graft | scion | rootstock | graft | scion | graft | scion | rootstock | graft |
| --- | --- | --- | --- | --- | --- | --- | --- | --- | --- | --- |
| Side grafting | | | Saddle grafting | | | Splice grafting | | Whip and tongue grafting | | |

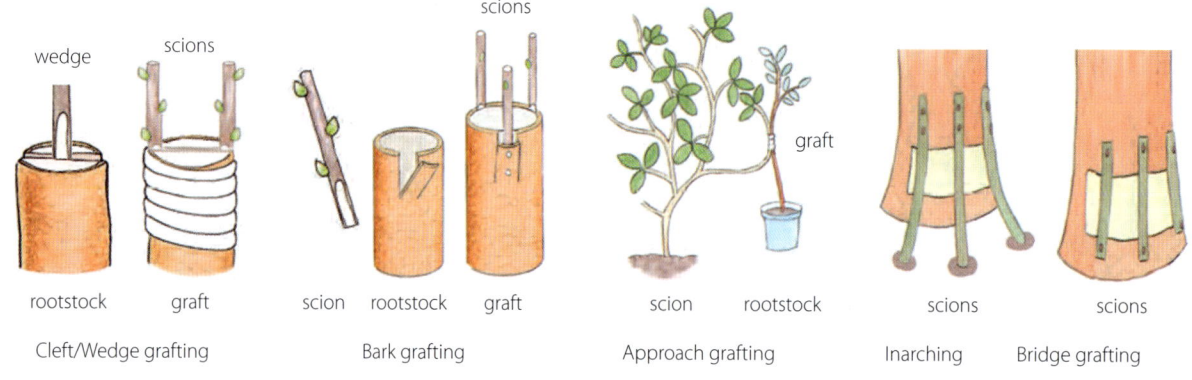

Cleft/Wedge grafting     Bark grafting     Approach grafting     Inarching    Bridge grafting

**granular fertiliser** *see* slow release fertilisers

**grasses**

Grasses include cereal grasses like wheat and barley, bamboos, natural and cultivated grasses, and pasture grasses for animals.

Grass stems (culms) are round and hollow or pithy except at the solid nodes. Leaves grow from the nodes and are usually long and narrow. Propagation is by seeds, and vegetatively by plugs, division of the crown, stolons, rhizomes, and separation of tillers.

Grasses have been adapted for all kinds of uses. Turf grass can be cut and sprouts again because the growing point (crown) is below the blades of a mower. Cereal crops and sugarcane, a perennial tropical grass, have been developed for food. Australian ornamental grasses are mostly naturally drought tolerant and require little water.

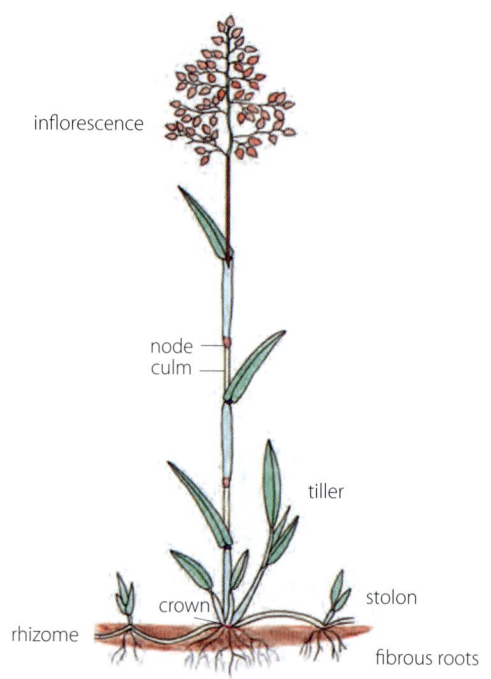

**green compost material** *see* compost

**green manure** *see* cover crops

**green-grafting**

Green-grafting allows growers to propagate cultivars after the time for dormant grafting has ended. It is done on some herbaceous and woody plants, and others after dormancy, when sap is flowing.

A rootstock that offers disease resistance and vigorous growth is the in-ground part of the plant. The scion, with desirable fruit qualities, is the above-ground part of the plant that is grafted onto the rootstock.

The earliest known references to green-grafting date back to the 5th century in China. The practice spread into East Asian countries, and by the early 20th century grafting of watermelon and cucumber was common in Japan. Grape vines are propagated with this method using top-grafting.

In addition, some plants of the same family, like potato and tomato, can be grafted together to create one plant. The rootstock will produce potatoes, and the scion will produce tomatoes.

## GREEN-GRAFTING TOMATO SEEDLINGS

trimmed

scion

rootstock

The seedlings for the rootstock and scion may be in one pot, or in separate pots then planted together after grafting

Incisions are made in the rootstock and the scion with a sterilised razor blade

The scion is gently inserted into the rootstock and clipped in place

The potted graft is placed in a plastic bag or other protection and kept in a humid dark place.

The scion and rootstock establish a connection in about seven days, and need a further seven days for the union to fully heal.

Plants then rest in a greenhouse for two or three days before hardening up and planting out.

## greenhouse

A greenhouse is a structure made of glass (a glass house), polycarbonate glazing, or plastic that provides protection for growing plants.

Greenhouses are warmed by the sun. If a greenhouse has artificial heating, it is called a hot house. A greenhouse and a hot house do not need to be separate structures. By equipping a greenhouse with heating, it can be transformed into a hot house when required.

## GREENHOUSE

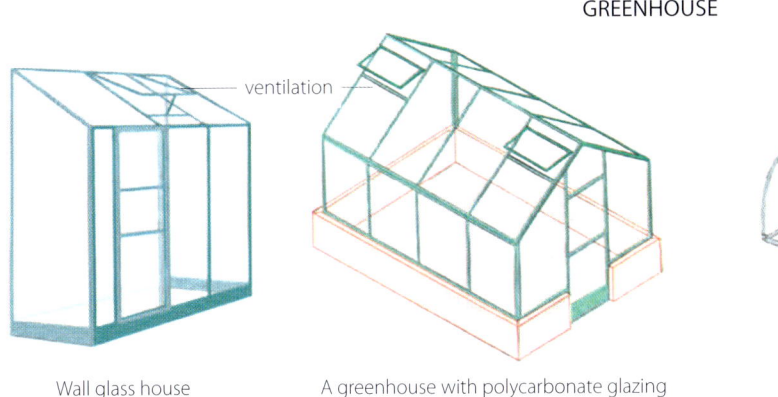

ventilation

Wall glass house

A greenhouse with polycarbonate glazing

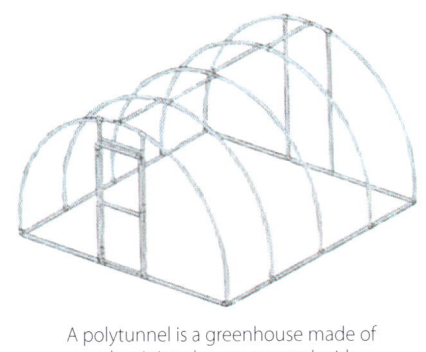

A polytunnel is a greenhouse made of aluminium hoops covered with UV-resistant polythene plastic

## greensand

Greensand is a powdered, ancient sedimentary rock that once lay beneath the sea.

The chemical remains of marine plant life in the rock are rich in nutrients that are released slowly and are easy for plants to use.

Greensand is not water-soluble and breaks down in the soil. The rock glauconite, from which it is made, gives it its bluish-green colour.

## greenwood cuttings

Greenwood cuttings are taken later than softwood cuttings, in the summer, when the stem is slightly harder and woodier. They are prepared in the same way as softwood cuttings.

## grey water

Untreated grey water is waste water from baths, showers, hand basins, laundry tubs and washing machines, that can be used for watering lawns or gardens. It is not contaminated with faeces.

Grey water does not include waste water from toilets or kitchens.

*see also* **blackwater, ground water**

## groundwater

Groundwater is water that has filtered down through the surface soil and saturated the empty spaces between rocks and material like gravel, sand and silt further underground.

The water table separates the water-saturated zone from the unsaturated zone above it.

The saturated zone forms an aquifer (an unconfined aquifer) that has an impenetrable layer beneath it but not above it. The layer prevents the stored water from seeping away. Water is pumped to the surface into wells or for distribution.

Sometimes there is an impenetrable layer above and below the aquifer that creates a confined aquifer. The pressure in the confined aquifer builds up, and the water from a drilled well will come to the surface without pumping, also called artesian water.

Thirty percent of the Earth's potable water is stored as groundwater. It supplies about 30% of all water used in Australia, including for maintaining gardens. Pumping too much water out of an aquifer reduces the flow and will eventually cause it to run dry. It can take many years, sometimes thousands, to replenish the water in an aquifer.

*see* **blackwater, grey water**

GROUNDWATER

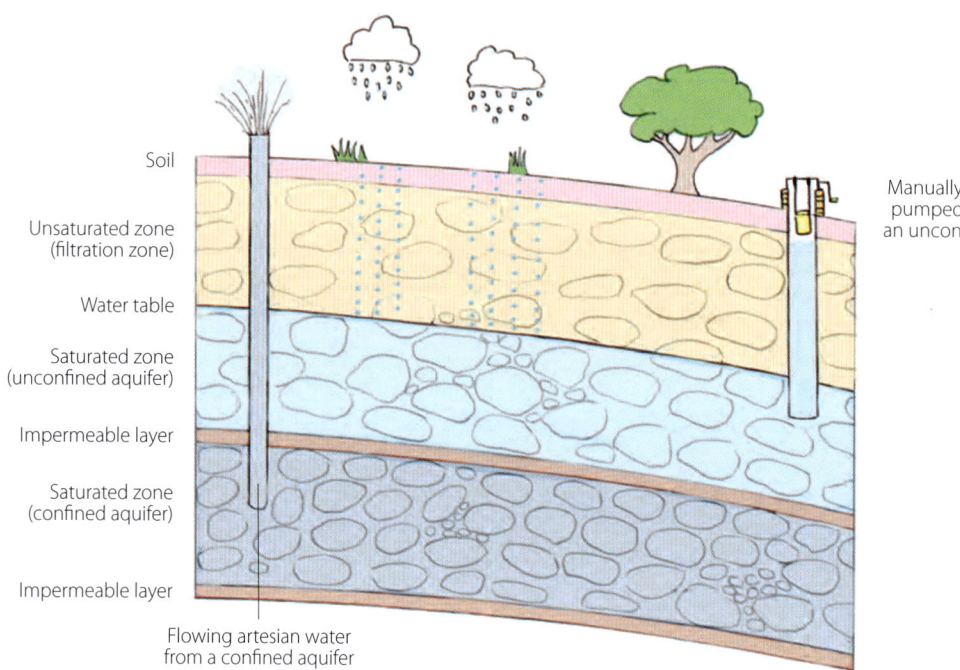

Soil

Unsaturated zone
(filtration zone)

Water table

Saturated zone
(unconfined aquifer)

Impermeable layer

Saturated zone
(confined aquifer)

Impermeable layer

Flowing artesian water
from a confined aquifer

Manually retrieved or
pumped water from
an unconfined aquifer

# guilds

Plants in nature grow naturally with other species. A permaculture guild is an artificial grouping of species that grow and support each other by building the soil, providing shade and ground cover, conserving water, making nutrients available, attracting beneficial insects, and repelling pests. All of this makes for a highly productive food space.

A guild has layers of plantings. It is built around a centrepiece that is an overstorey plant, usually a tall fruit or nut tree. Below this are understorey plants like citrus, dwarf fruit trees and hazelnut. Then there are shrubs, herbaceous plants, ground covers, vines and roots.

Each plant in the guild performs a valuable function. Accumulators draw nutrients from the soil and concentrate them in their tissue. Chicory accumulates minerals such as calcium and potassium. They are returned to the soil when chicory mulch rots down. Mulches, like comfrey, that are cut and dropped on the soil, prevent evaporation and also return nutrients to the soil. Fixers are usually legumes, like alfalfa that has nitrogen-fixing nodules on its roots. Suppressors, like bulbs, prevent the spread of weeds. Attractors bring in insect pollinators. Repellers deter pests.

Guilds may not have all of the above elements. There is no specific formula for a guild, and they can be designed for gardens of any size.

Black walnut can be the centrepiece of a guild. However, its roots exude a chemical called juglone, that inhibits the growth of many other plant species. Some plants that are useful for growing with walnuts are mulberries, currants, gooseberries and raspberries.

APPLE TREE GUILD

Overstorey and centrepiece
apple tree
(*Malus domestica*)

Accumulator
chicory
(*Cichorium intybus*)

Understorey tree
hazelnut tree
(*Corylus avellana*)

Suppressor
garlic
(*Allium sativum*)

Insect attractant
prostrate thyme
(*Thymus serpyllum*)

Insect repellent
nasturtium
(*Tropaeolum majus*)

Understorey tree
orange tree
(*Citrus x sinensis*)

Herbaceous plant
Nitrogen fixer
alfalfa
(*Medicago sativa*)

Mulch
comfrey
(*Symphytum officinale*)

Shrub
raspberry
(*Rubus idaeus*)

## gummosis

Gummosis is the oozing of gum from a wound on a woody plant. The sticky gum hardens and covers the wound.

The cause may be an injury or the result of an attack by insects or pathogens.

oozing gum

## gymnosperms

There are two groups of seed-bearing plants: gymnosperms, that lack flowers, and angiosperms, that have flowers.

Gymnosperms are woody trees and shrubs. Their reproductive structure is usually a cone. Female cones bear ovules and seeds exposed on scales, and male cones bear pollen, also exposed in a sac on a cone. Gymnosperm seeds have more than two seed leaves (cotyledons).

There are four divisions of gymnosperms: conifers, cycads, ginkgos, and gnetales. Conifers, like pine, spruce and larch, are usually needle-leaved evergreens. Cycads are tropical plants, with a single thick stem, used as ornamentals. There is only one ginkgo, *Ginkgo biloba*. It has fan-shaped leaves and is widely planted. Gnetales are an unusual group of three genera, *Ephedra*, *Gnetum* and *Welwitschia*, that are little known in horticulture.

GYMNOSPERMS

cotyledons

pine (*pinus*)

Gymnosperm seeds have more than two seed leaves (cotyledons)

male cone    female cone

pine (*Pinus*)

winged seeds on female cone scale

Conifer

Cycad

## gypsum

Gypsum is a naturally occurring mineral. It is suggested as an amendment for breaking up sodic clay soils. A soil test should be done on clay soil to see that it is sodic before applying gypsum.

## habit

Habit refers to the characteristic shape and growth form of a plant, such as a herb, vine, shrub or tree. It includes its size and colour.

Terms like erect, creeping, woody, tufted etc. further describe the habit.

Climbing liana      Creeping herb      Erect tree

## habitat

Habitat refers to the natural environment in which plants or other organisms live, such as a grassland, woodland or rainforest. It includes climate and soil.

A gardener who knows the original habitat of a plant will have an understanding of whether or not it is a good choice for local growing conditions.

## half-hardy

A half-hardy plant can cope with temperatures down to 0°C (32°F) but below that needs protection from frost.

## hand pollination

Hand pollination is the manual transfer of pollen from the anthers of the male part of a flower (stamen) to the stigma of the female part of a flower (pistil).

Pumpkins carry male and female flowers on the same plant. Flowers are pollinated by nectar-collecting bees. Pumpkins can be hand-pollinated by using a small brush to transfer the pollen from the anthers of the male flower to the stigma of the female flower.

Tomatoes have the pistil enclosed by the stamens. They are self-fertile, and shaking the flower is enough to transfer the pollen to the stigma.

HAND POLLINATION

three yellow stigmas

female flower

male flower

anthers

Flowers can be hand pollinated by transferring pollen from the male anthers to the female stigmas with a small paint brush.

**Unisexual flowers** of pumpkin (*Cucurbita pepito*)

stigma

fused anthers

Pollen is transferred to the stigma inside the fused anthers. Shaking the flower will transfer pollen to the stigma.

**Bisexual flower** of tomato (*Solanum lycopersicum*)

## Hanging Gardens of Babylon

Of the Seven Wonders of the Ancient World, the Hanging Gardens of Babylon is the only wonder for which no trace of its existence has been found. They are said to have existed in Babylon between 605 and 562 BC.

There is no mention of them in contemporary inscriptions, though they are discussed much later in Greek and Roman sources.

## hardening off

Hardening off, also called acclimatisation, is the process of preparing seedlings, cuttings or plants grown inside a greenhouse or hot house to grow outside in the garden.

A sheltered space or a cold frame allows the plant to adapt to changes in temperature and humidity while it is protected from the weather. In seedlings, it encourages slower, more sturdy growth.

## hardiness

Hardiness is the measure of how well a plant can survive in different climates.

## hardpan

Hardpan is a hardened, compacted or cemented layer of soil in the soil profile.

No water can penetrate the soil beyond the hardpan, and roots cannot break through it. It can be a natural occurrence or human-made.

Digging and turning over soil exposes it to the ultraviolet rays of the sun that sterilise the soil and kill soil organisms. Soil loses a lot of its nutrients and organic matter and doesn't retain water well. This destroys the soil structure and leads to hardpan formation.

Garden soils that have previously been commercially farmed may have this problem.

## hardwood cuttings

A hardwood cutting is a piece of stem at least as thick as a pencil and 25–30 cm long, cut from the current year's growth that has turned hard and woody.

Evergreen plants like camellia, cedar and rhododendron, and deciduous plants like crepe myrtle, flowering quince and roses, can be grown from hardwood cuttings. Hardwood cuttings from deciduous trees are taken after leaf fall, from mid-autumn to late winter. However, evergreen plants are usually propagated from tip cuttings

Hardwood cuttings are often wounded to improve root development. They can be propagated in the ground or in pots.

HARDWOOD CUTTINGS

softwood

semi-hardwood

hardwood

rooting hormone

Lower leaves are removed

Large leaves are cut back to about half to reduce transpiration

Bark removed (wounded) to encourange rooting

mallet cutting    heel cutting    simple cutting

tip cutting

**Deciduous hardwood cuttings**    **Evergreen hardwood cutting**

## hardy

A hardy plant can survive throughout the year in local conditions like drought, extreme heat or cold, and poor soil.

## heading back

Heading back is a method of pruning the terminal branches to control tree size and encourage lateral growth and bushiness.

The ends of a woody branch are cut back to 1 cm above a bud that faces in the direction you want new growth to occur. New shoots will develop from these buds.

**heap composting** *see* compost

**heartwood** *see* xylem

## heat mat

A heat mat is an electrically heated mat that is placed under a propagating tray or pots to gently warm the soil and promote faster germination and strong, healthy seedlings and cuttings.

## hedge

A closely planted row of shrubs made up of a single species.

## hedgerow

A strip of land densely planted with a variety of perennial species.

## hedging

Hedging is trimming a row of plants with manual or battery shears. The aim is to make a formal hedge with a geometrical shape or a more natural, loosely clipped informal hedge.

The process encourages leaf growth on the outer surface of the hedge and fewer leaves in the interior.

The interior branching may need to be thinned out every few years to allow light to enter and prevent disease.

Informal lavender hedge or border

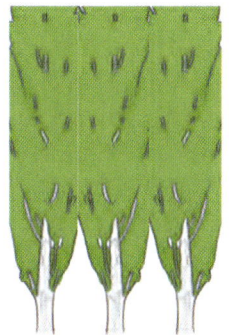

Formal poplar hedge

## heel cutting

A side-shoot (cutting) is gently peeled away from the parent stem, together with a piece of bark or wood (the heel). This exposes more of the layer of cells that will grow into roots.

The heel is dipped in rooting hormone before planting to encourage root growth.

parent stem
cutting
heel

Heel cuttings are used to propagate semi-hardwood and hardwood cuttings

## heeling in

Heeling in is used to store bare-rooted plants for short periods of time if they can't be planted immediately.

The plant is placed in a V-shaped trench so that it leans against the slope, with the roots in the trench and the rest of the plant extending out of it. The roots are then covered with loose soil, compost or wood shavings, forming a slight mound at the top. The roots are kept moist but not wet.

loose soil, compost or wood shavings

V-shaped trench

### heirloom

Heirloom plants are understood to grow from seeds handed down from one generation to the next over time.

Growing heirlooms is a way of saving genetically diverse plants that have been bred locally over many years. Heirloom plants are open-pollinated, that is, they are pollinated naturally. Their seed produces plants that remain fairly consistent from one generation to the next.

There are many heirloom carrot varieties, including some that have been around for hundreds of years and are still available as seed.

*cf.* **hybrid**

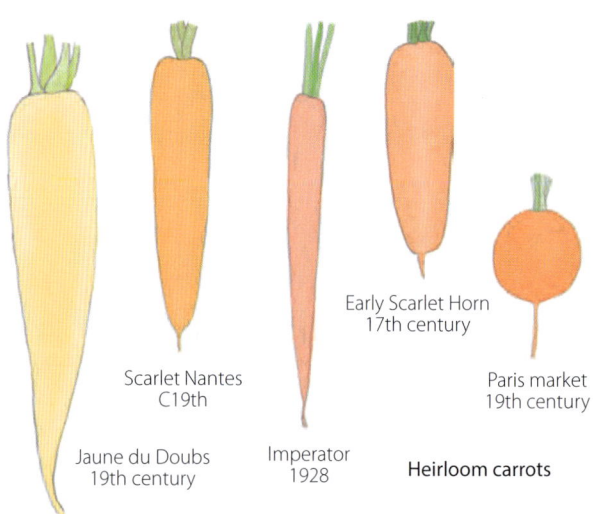

Scarlet Nantes
C19th

Early Scarlet Horn
17th century

Paris market
19th century

Jaune du Doubs
19th century

Imperator
1928

**Heirloom carrots**

### hen and chicken fern

The hen and chicken fern (*Asplenium bulbiferum*) develops tiny plantlets called bulbils on the fronds. They will take root easily on moist soil.

The succulent *Echeveria* is similarly called hen and chicks.

hen and chicken fern
(*Asplenium bulbiferum*)

bulbil

### herb

A herb is a plant that does not develop a woody stem but has soft stems and is usually green and soft in texture. It may be annual, biennial or perennial, and includes clover and many garden flowers like sunflowers and daylilies. Grasses are herbs.

A herb also describes a subset of vegetables that are grown in a garden for flavour (such as sage, basil and coriander), medicinal use (chamomile and *Echinacea*) and fragrance (lavender and scented pelargoniums).

*see also* **herb gardens, medieval gardens**

### herb gardens

Botanically, a herb is a plant that does not develop a woody stem and is usually green and soft in texture. It is annual, biennial or perennial. Culinary herbs, medicinal herbs, and herbs valued for fragrance and dyeing, however, can be herbaceous (non-woody) like basil, or woody shrubs like rosemary and lavender. Bay leaves, from the laurel tree, are used as a herb.

Ancient civilisations like those of China, Egypt and Sumeria used herbs to treat illness as well as to flavour food and make perfumes.

In medieval Europe, monastery herb gardens (physic gardens) were the pharmacies of the day. Monks in Europe and England learnt from ancient Greek, Roman and Arabic herbals and studied plants that were not understood but were thought to have medicinal potential. Remedies of blended herbs or single herbs (simples) were used to treat patients being cared for in a monastery.

These well-organised gardens for growing and studying medicinal herbs, were precursors of botanic gardens. The first universities established in Europe in the 16th century taught medicine and had teaching gardens of medicinal herbs.

**HERBS FOR FLAVOUR, MEDICINE, ESSENTIAL OILS AND DYES**

Thyme, sage, rosemary and mint are essential ingredients in cooking around the world

Sage (*Salvia officinalis*) after a 15th century copy of '*Liber de Simplici Medicina*' by Platearius, late 12th century

The stigmas of saffron (*Crocus sativa*) were used as a yellow dye in ancient Persia and, later, to colour the robes of Buddhist monks in Asia

Saffron crocus (*Crocus sativa*) after '*A Curious Herbal*', by Elizabeth Blackwell, 1737

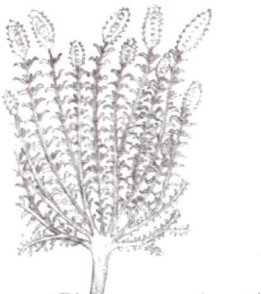

Lavender (*Lavandula*) and lily of the valley (*Convallaria majalis*) are distilled to make essential oils for fragrances

Lavender (*Lavandula stoechas*)) after '*Erbario Greco*', 50–70 AD by Dioscorides, and lily of the valley (*Convallaria majalis*), after a 15th century copy of '*Liber de Simplici Medicina*' by Platearius

Cannabis (*Cannabis sativa*) and chamomile (*Chamomilla recutita*) have been used throughout time for medicinal purposes

Cannabis (*Cannabis sativa*), after '*De Stirpium*' by Hieronymus Bock, 1546, and chamomile (*Chamomilla recutita*), detail after '*De Historia Stirpium*' by Fuchs, 1542

## herbaceous cuttings

Herbaceous cuttings are made from the new, soft green growth of non-woody, herbaceous plants, like chrysanthemums and dahlias. They are prepared in the same way as softwood cuttings.

## herbicide

A herbicide is a chemical used to kill undesirable plants. A selective herbicide will kill or injure some plants without harming others. A broad-spectrum herbicide will kill all plants that it comes in contact with.

## honeydew

Honeydew is a sweet, sticky liquid excreted by plant-sucking insects like scale, mealy bugs, aphids and white flies.

The honeydew drips onto stems and leaves and attracts the sooty mould fungus. As the mould grows, it may block out enough light to inhibit photosynthesis. Some ants feed on the honeydew of aphids. They stroke their antennae to coax them to produce honeydew. In return, the ants tend the aphids by keeping predators away.

## horticultural fleece

Horticultural fleece is a soft, lightweight polyester material that is porous and allows water and air through.

It is used to provide protection against cold and frost, and as a cover to warm soil before sowing or planting out. It is used outside and in unheated greenhouses and polytunnels as a way of extending the growing season. Horticultural fleece will protect against pests, like cabbage white butterfly that will be prevented from laying its eggs on brassicas.

Horticultural fleece comes in different weights, widths and lengths. The weight controls light transmission and shade. Heavier weights can be protective at night and removed during the day.

**hot composting** *see* compost

**hot house** *see* greenhouse

## hotbed

A hotbed is a cold frame heated by electricity or a deep layer of decomposing organic matter like manure and straw under a layer of loam.

It is used to extend the sowing and growing season in cool months and can be set up outside or in a greenhouse. Depending on the climate, a manure-heated bed will last several weeks. The decomposed organic matter can be used as compost.

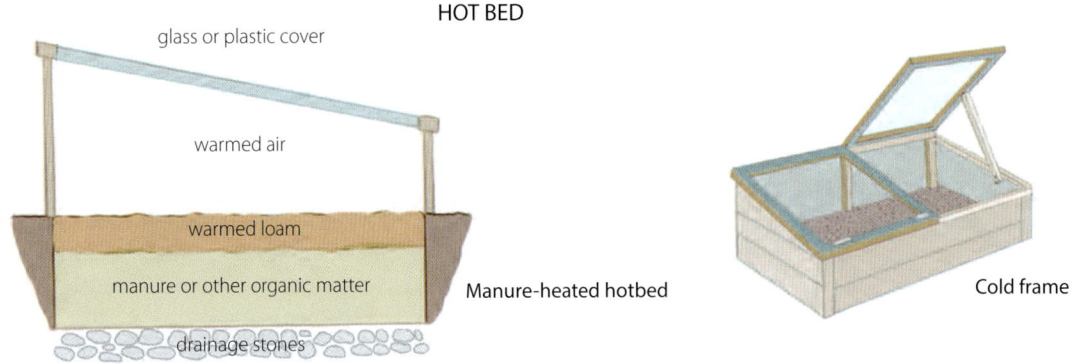

HOT BED

glass or plastic cover

warmed air

warmed loam

manure or other organic matter

Manure-heated hotbed

drainage stones

Cold frame

**hugelkulture** *see* next page

## humidity

Humidity refers to the amount of moisture in the atmosphere.

## humus

Humus is organic matter that has decomposed further than compost. Organisms can't easily break it down, so it accumulates in the soil. However, plant roots exude a mild acid that breaks it up. The nutrients are washed into the soil moisture and are absorbed by the roots.

Humus not only holds nutrients, but it also swells when wet, making it very good for retaining soil moisture. Humus is a store of carbon in the earth. If there is insufficient drainage and oxygen, as in some wetlands, it becomes peat.

## hybrid, hybridisation

A hybrid seed results from cross-pollination between two different species or genera.

The plant that bears the seed is called the female or seed parent, and the plant that provides the pollen to fertilise the female, is called the male or pollen parent. The resulting seed produces a hybrid plant with characteristics of both parents. Hybridisation does occur in nature, but is mostly done by humans.

A hybrid plant is indicated by the genus name and species name having an x between them. The sweet orange (*Citrus* x *sinensis*) is a hybrid between a pomelo (*Citrus maxima*) and a mandarin (*Citrus reticulata*).

Some hybrid seeds grow true to type. Usually, however, hybrids must be propagated vegetatively, as by grafting, cuttings and tissue culture. A hybrid itself can be used in cross-pollination to produce a new hybrid.

**hybrid vigour** *see* F1 hybrid

## hugelkultur

A mounded garden bed with large wooden logs at the base and, on top of that, layers of smaller logs, branches and twigs. These are topped with composting materials that include carbon browns like straw, and nitrogen greens like grass clippings and manures. The alternating layers can include mulch, and are finally topped with soil.

It is possible to plant out your garden straight away. Soil aeration increases as the branches and logs break down so that the bed will not need to be tilled.

Hugelkultur beds are self-fertilising and act like slow-release compost piles that feed the microbial life in the soil.

The buried logs slowly decompose into humus that feeds plants from below. The spongy composition of the humus helps retain water and holds it for long periods in the soil, where it can't evaporate and is available to plants in drier times. The carbon in the logs remains sequestered in the soil.

The name comes from the German for 'mound culture'. The system has been used in Germany and Eastern Europe for hundreds of years and is now popular in permaculture.

HUGELKULTUR

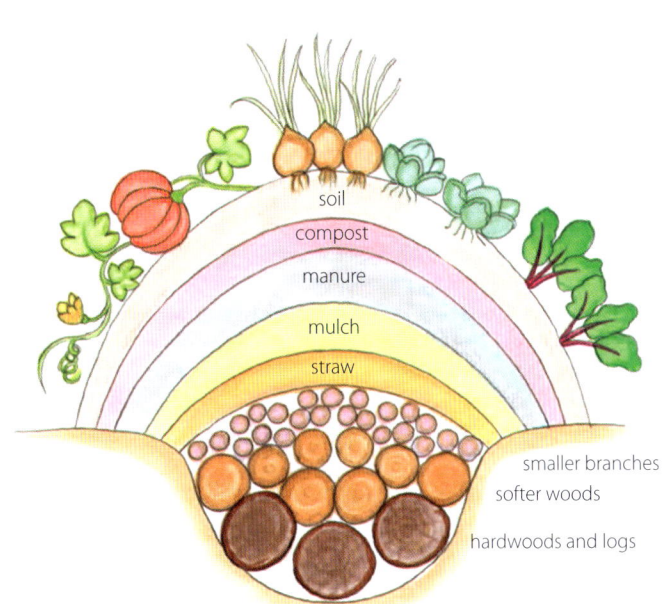

soil
compost
manure
mulch
straw
smaller branches
softer woods
hardwoods and logs

## hydrophilic *see* soil wettability

## hydrophobic *see* soil wettability

## hydroponics

In hydroponics plants have roots growing in nutrient-rich water instead of soil.

Water use is reduced significantly, making the system suitable for growing crops in arid areas. Hydroponics works well for urban farming.

Systems can vary greatly. All have plants supported in the water with or without an anchoring plug to hold the plant. Plugs are made of a neutral medium, like vermiculite or coco coir, that is held in a small container.

Pumps circulate and recycle the water through the system. Another pump increases the oxygen content in the nutrient solution by forcing air through an air stone, much like that found in fish tanks.

Plants also need natural or artificial light and stable air temperature and humidity. Systems can be built to allow for the vertical growth of climbers like beans and tomatoes.

**HYDROPONICS**

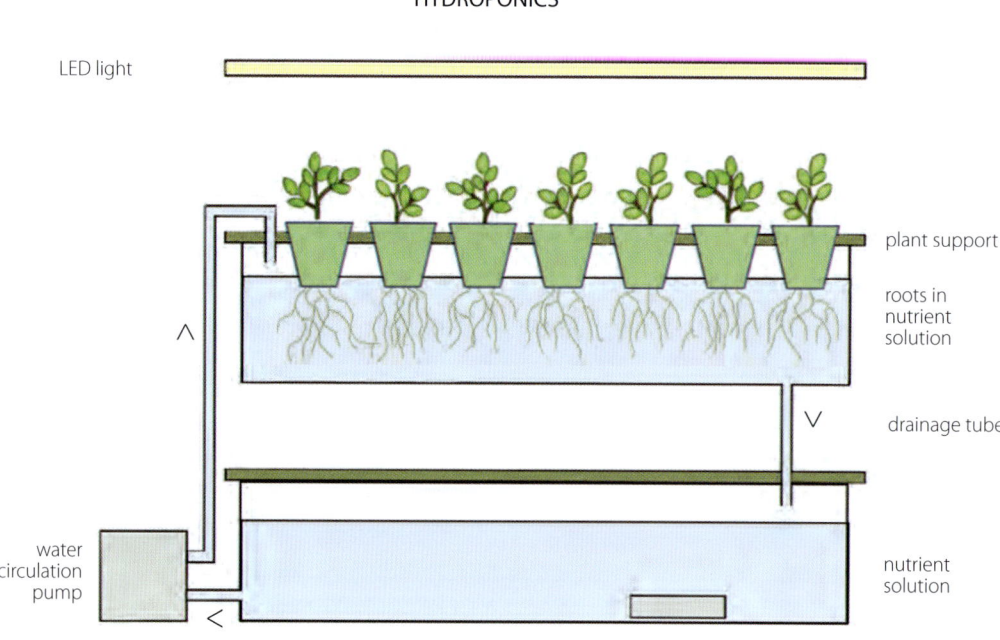

LED light

plant support

roots in nutrient solution

drainage tube

water circulation pump

nutrient solution

air stone

---

**hypogeal germination** *see* germination

**imbricate bulbs**

Imbricate bulbs have fleshy overlapping scales that are called 'naked' because they lack a papery tunic, like that on an onion.

The fleshy scales store nutrients and water for the bulb when it starts to grow. The leafy flower stalk in the centre of the bulb arises from the basal plate, as do the roots and the scales. True lilies in the genus *Lilium* have imbricate bulbs.

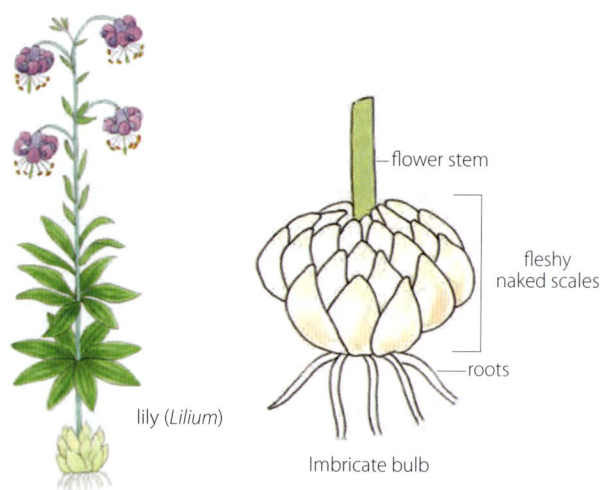

flower stem

fleshy naked scales

roots

lily (*Lilium*)

Imbricate bulb

**imperfect flower**

A unisexual flower that has only functioning stamens or functioning pistils.

**inarching**

Inarching is used to repair the severed root system of a damaged tree.

The damaged area is trimmed and cleaned. Small, compatible seedlings with flexible stems are planted around the damaged area to use as rootstock.

Vertical cuts are made through the upper edge of the damaged trunk area and sliced away. The edge of the seedling on the side facing the tree is cut away to match the cut on the trunk. They are then bent inward and nailed to the undamaged part of the tree.

INARCHING

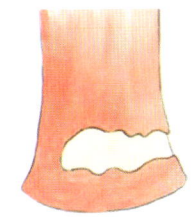

1. Nutrients and water cannot move across the damaged area on the tree

2. Seedlings are planted at intervals around the base of the damaged trunk

3. Seedlings are cut on the side facing the tree. The cut area matches that on the trunk.

4. The grafted roots of the seedlings will replace the severed tree roots

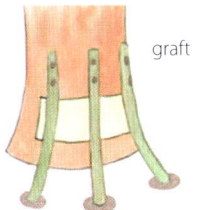

graft

5. The scions are grafted onto the undamaged tree trunk. Nutrients and water flow through them to the tree.

---

## incomplete flower

An incomplete flower naturally lacks one or more of the four whorls of sepals, petals, stamens, and/or pistils.

## indeterminate growth

Indeterminate growth refers to a flowering (and fruiting) stem that will grow at the tip indefinitely.

Flowers at the base of the shoot form flowers (and fruit) first. Flowering and fruiting will continue up the stem indefinitely. This provides an ongoing crop that can be harvested multiple times.

Plants with indeterminate growth include vining cucumbers that use tendrils to climb, climbers like tomatoes that need to be staked, and pole beans that climb by twining their stems around a support.

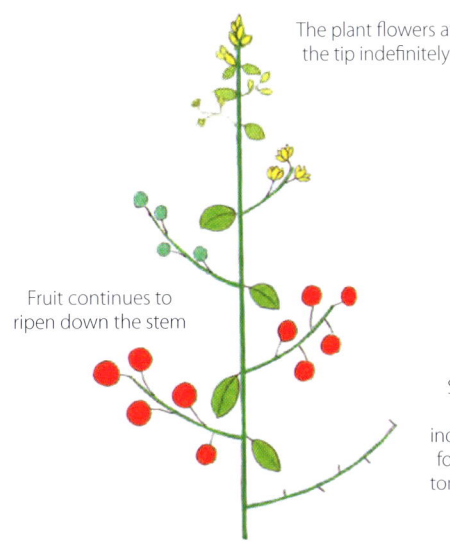

The plant flowers at the tip indefinitely

Fruit continues to ripen down the stem

Some plants have determinate and indeterminate growth forms. These include tomatoes, cucumbers, peas and beans

## indigenous gardens = native gardens

## indoor plants

Indoor plants are suited to growing inside in pots, for their foliage or flowers.

Each has particular needs for light, temperature, humidity, water and fertilisation. The trailing vine pothos (*Epipremnum aureum*) will grow in soil or water in almost any environment. The moth orchid (*Phalaenopsis*) is long-flowering and long-lived. Peace lilies (*Spathiphyllum*) have beautiful leaves all year, and with enough light will flower over a long period.

Cyclamens flower in winter and die back in spring.

Watering is reduced during this dormant period until the above-ground tuber-like structure begins to shoot in autumn.

## inflorescence

An inflorescence is the arrangement of a cluster of flowers on a plant. It includes its stems, stalks, bracts, bracteoles and flowers.

The main stem holding the whole inflorescence is the peduncle, and smaller side stems are pedicels.

The purpose of the inflorescence is to attract pollinators and make seeds that will reproduce the plant. The male pollen from a flower fertilises a female egg in an ovary. This will develop into a seed.

INFLORESCENCE

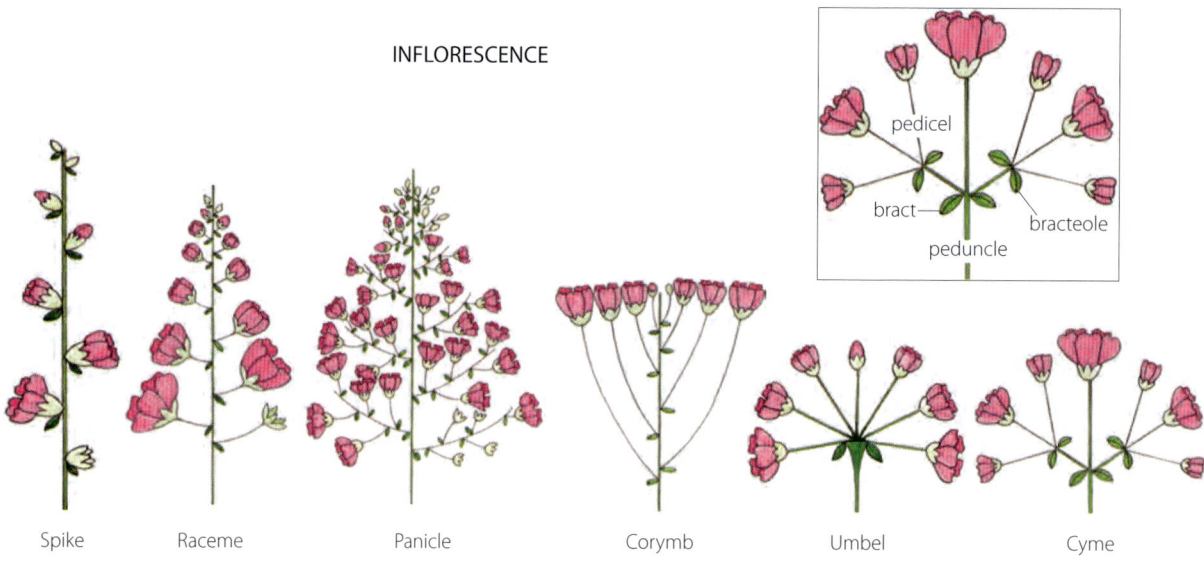

Spike    Raceme    Panicle    Corymb    Umbel    Cyme

---

**inner bark** = phloem

## insecticide

An insecticide is a substance that kills insects.

*see also* **pesticides, pyrethrum**

**intercropping** = interplanting

**internode** *see* stem

**interplanting** *see* next page

## interstock

In grafting, an interstock is used when the scion and rootstock are incompatible.

The interstock is compatible with both the rootstock and the scion, and is grafted between the two to join them together.

This process is called double working.

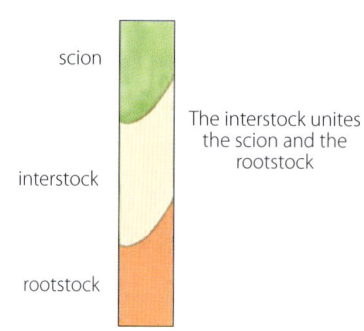

scion

interstock

rootstock

The interstock unites the scion and the rootstock

## interplanting

Interplanting, also called intercropping, is a method of planting a fast-growing crop between a slower-growing crop at the same time or a mixture of crops in the same space.

An example is Brussels sprouts, that take 80–90 days to mature, interplanted with spinach, that takes 42 days to mature. Spinach can be harvested as needed while acting as a living mulch that keeps soil moist and suppresses weeds while the sprouts are growing.

The plantings can be in rows or mixed. The traditional Three Sisters Garden has plantings of corn, beans, and pumpkin, with each plant providing a benefit to the other. Companion planting is another form of interplanting where beneficial flowers like marigolds are planted with cabbages and tomatoes.

Orchards are interplanted while the trees are growing and once they are established.

Corn supports climbing beans

Beans add nitrogen to soil

Squash and pumpkin provide living mulch

Three Sisters Garden

A row of slow-growing Brussels sprouts interplanted with fast growing spinach

## introduced plants *see* exotic gardening

## iron *see* plant nutrients

## irrigation

The aim of irrigation is to provide plants with water, so they can develop vigorous root systems for extracting nutrients from the soil.

Sprinklers and drip irrigators are popular in gardens.

Sprinklers spray water overhead, and drip irrigation supplies water slowly and specifically to the roots. Both systems use different kinds of nozzles attached to pipes and tubes that supply water to the garden space. Sprinklers have nozzles that apply water in a specific pattern and at a chosen distance. Nozzles can be changed and the flow adjusted to suit the season and size of the plant as it grows.

There are also soaker hoses that have holes along their length to release water into the soil. Irrigation systems can be turned on and off manually, or automatically with timers. Most gardeners use hand-held hoses and watering cans.

Water needs of different crops are considered when setting up an irrigation system. Vegetables with shallow root systems, like cabbage and lettuce, need to be watered frequently. Deep-rooted crops, like tomatoes, require a longer and less frequent watering. Some vegetables, like beans and melons, are watered at the base to prevent leaf diseases, and others, like carrots and turnips, can be watered overhead.

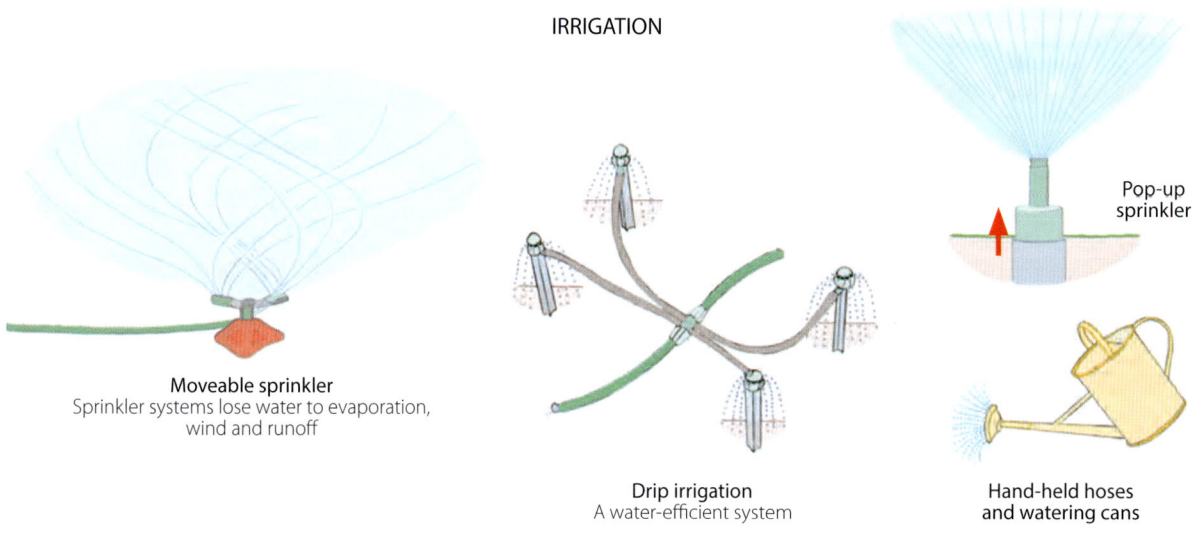

IRRIGATION

**Moveable sprinkler**
Sprinkler systems lose water to evaporation, wind and runoff

Pop-up sprinkler

**Drip irrigation**
A water-efficient system

**Hand-held hoses and watering cans**

## Islamic gardens *see* page 77

Islamic gardens *see* page 77

## Japanese gardens

A Japanese garden reflects the beauty of nature. It is a space away from the busyness of the real world that invites peaceful contemplation. Features in the garden have a deeper, symbolic meaning. A rock can represent a small island or a mountain; water symbolises purity; and certain plants, like pines and evergreens, indicate permanence.

There are different kinds of Japanese gardens. The dry garden, *karesansui,* also known as a Zen garden, represents natural scenery with rocks and carefully raked gravel, but no water. The tea garden, *roji,* is a simple rustic garden with stepping stones that provide a meditative approach to a tea house. The courtyard garden, *tsuboniwa,* is a garden within the walls of a residence or an enclosed space beside it.

Many gardens have a stone water basin, *tsukubai,* usually with a bamboo dipper for scooping up water. These basins are used for ritual cleansing in tea gardens and are often paired with a lantern.

Trees and shrubs are trimmed and pruned to reflect their natural shape and to maintain their size to suit the proportions of the garden. Flowers are not a feature, but moss is valued. Cloud pruning has a special place in Japanese gardens.

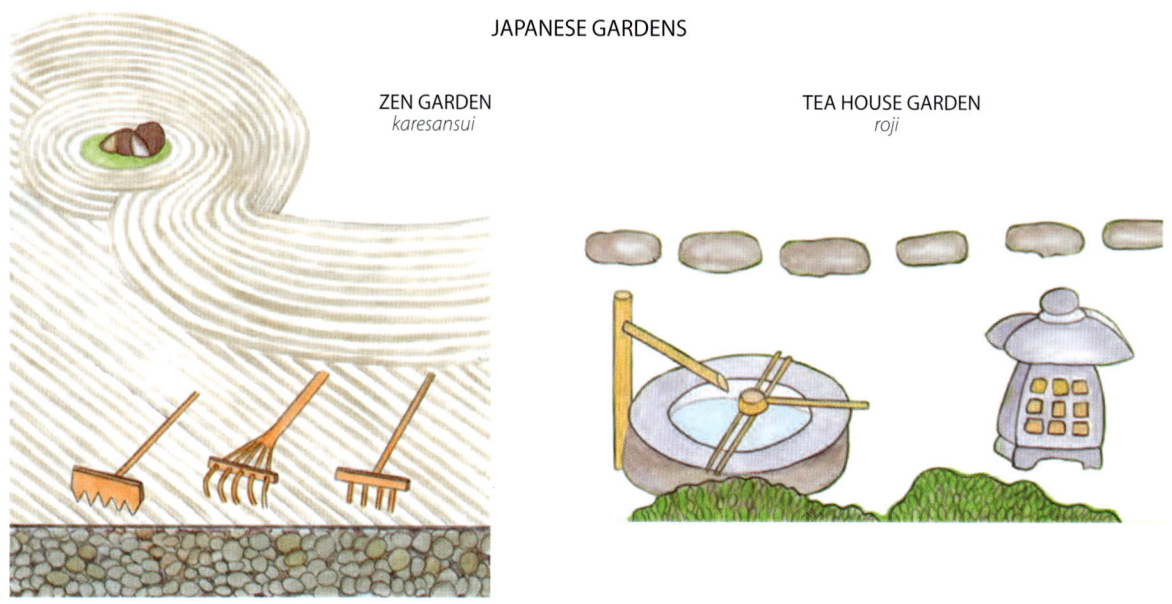

JAPANESE GARDENS

**ZEN GARDEN**
*karesansui*

**TEA HOUSE GARDEN**
*roji*

## Islamic gardens

Islamic gardens began in Saudi Arabia with the rise of Islam in the 7th century. As Islamic civilisation spread, it dramatically changed the landscape in North Africa, the Iberian Peninsula, Turkey, and in the east, Persia (Iran), and India under the Moghuls. Islamic gardens have a deep spiritual significance and are seen as a place of rest and reflection, and a reminder of the heavenly Garden of Paradise.

A network of experimental gardens linked the Islamic world. Research was done on soils, fertilisers, adapting plants to different climates, grafting, and cross-breeding to produce new varieties. The Arabs introduced plants wherever they went, including citrus varieties, date palms, pomegranates, and almond trees into Europe.

Gardens provided a haven in the harsh desert surroundings. A house and garden, whether modest or grand, were surrounded by walls, and always featured water. The basic, symmetrical quadrangular or rectangular chahar bagh (four gardens) design was reproduced everywhere. It had fruit trees, fragrant flowers and shrubs, and was a place for contemplation. Palm trees, cypresses and cedars provided shade. It came to represent paradise on earth.

Sophisticated techniques, like qanats and reservoirs, were used to collect water. Complex irrigation systems, using gravity, channelled water into arid or unproductive landscapes. Orchards, vineyards, olive groves and vegetable gardens flourished.

The pinnacle of the grand Islamic garden is represented by Mughal creations, from Babur's gardens in Kabul to the Taj Mahal in India.

*see also* **Al-Andalus**

**A MOGHUL CHAHAR BAGH GARDEN**

Well with
noria (waterwheel)

Species from other parts of the world were prized. This could be a bottle palm (*Hyophorbe lagenicaulis*), native to an island near Mauritius.

Fruit trees included mangos, oranges and pomegranates

A symmetrical garden of four squares or rectangles divided by paths and waterways

The irrigation system relied on complex calculations that enabled water to be distributed by gravity

The Italian cypress (*Cupressus sempervirens*) was made popular in India by the Mughuls

After a miniature of a Moghul garden, the Bagh-e Yafa (Garden of Fidelity), built 1508–1509 near Kabul

## keiki

A keiki is a plantlet or offshoot that grows from the base of some orchids, or from one of the nodes on the stem.

It is a clone of the mother plant and can be detached and propagated once the rootlets have developed.

keiki

dendrobium (*Dendrobium*)

## keyhole gardens

A keyhole garden is a round garden bed about 2 m wide, with a notch in the side. In the centre is a basket made of wire mesh for composting materials, grey water and manure. This will rot down and feed the garden.

There is a foundation of drainage material like stones and twigs under the bed and the basket.

Cardboard, then green organic matter and brown organic matter are built up on the garden itself and topped with topsoil and mulch.

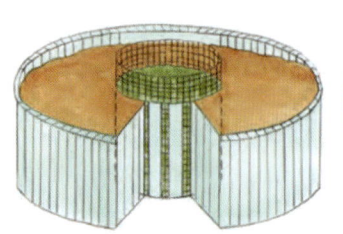

The garden and composting basket are easily accessible

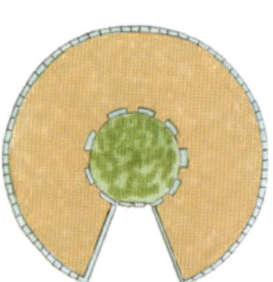

Preferred crops are leafy vegetables like lettuce and spinach, herbs, and root vegetables like carrot, onion, beetroot and turnip

## knot garden

A knot garden is a geometrical pattern of low hedges, often enclosed in a square, that are trimmed so that they appear to be woven above and below one another.

Different varieties of hedging plants are used to give the effect of different colours threading through the pattern. The spaces are usually filled with gravel rather than flowers.

Hedging plants traditionally include box, rosemary, lavender and southernwood.

The knot garden and the parterre together inspired herb garden designs.

C17th design for a knot garden by William Lawson

## labyrinth

A labyrinth is a simple, winding one-way path. It is walked in a spirit of meditation as a way to achieve calm and serenity. Hedges define the path and are low enough for the person walking through to see the general layout.

Labyrinths are found worldwide and date back to ancient times. There are simple plans to suit smaller gardens.

The garden around the paths is planted with low shrubs like lavender and box

*cf.* maze

**lasagne composting** *see* sheet composting

**lateral bud** *see* bud

## lawn

An area of short regularly mown, fertilised and watered area of grass. Gentler more eco-friendly options are gradually replacing this landscaping feature.

## layering

In layering, new plantlets form while attached to and nourished by the parent plant. A stem is buried in the soil and forms roots and shoots. Once established, the plantlet is separated and grown as an independent plant.

Layering can occur naturally when a branchlet bends and touches the ground, or the tip of the stem, such as a bramble, touches the ground.

When done artificially, a small patch of the area to be layered is often wounded. It is then coated with rooting hormone to encourage root formation before being buried. Some plants, like brambles that shoot from the tip, prefer to have only the tip buried.

PROPAGATION BY LAYERING

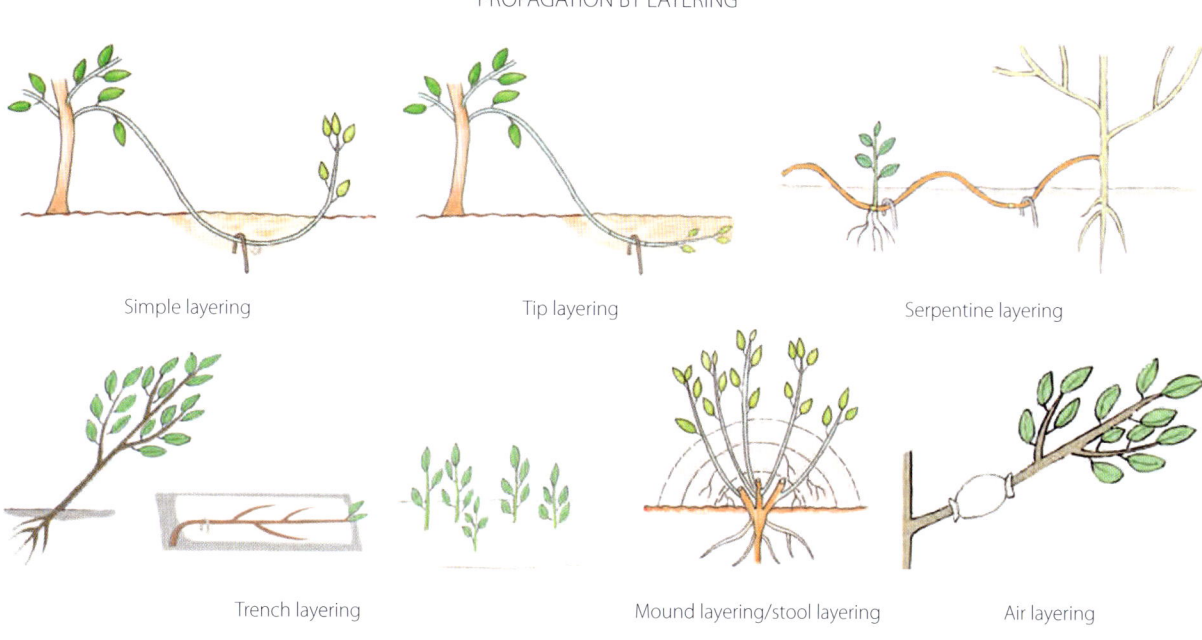

Simple layering                    Tip layering                    Serpentine layering

Trench layering          Mound layering/stool layering          Air layering

## leaching

Leaching is the loss of soluble plant nutrients from the topsoil due to rain and irrigation. If nitrate is leached, for example, the soil becomes acidic, which is detrimental to plant growth.

## leader

A leader is the dominant shoot of a tree or shrub, usually found at the tip of the whole plant.

If the central axis dies or is damaged, a new leader, called a lateral leader, may form a side branch.

Some plants have no dominant leader.

leader

lateral leader

## leaf

A leaf arises from a node on a stem and comes in many shapes and sizes. It usually consists of a flat blade on a stem-like petiole.

The surface captures energy from the sun for photosynthesis by means of chlorophyll. Chlorophyll gives the leaf its green colour.

Bracts, tendrils and spines are modified leaves.

Stamens, pistils, petals and sepals are also considered to be modified leaves.

*see also* **cotyledon**

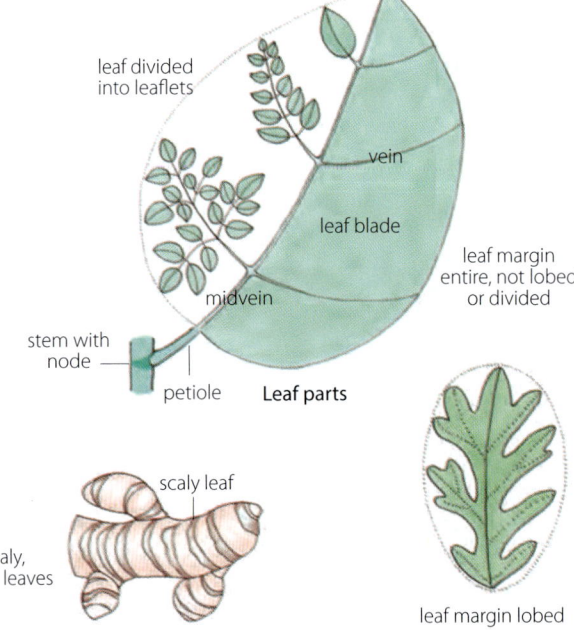

leaf divided into leaflets

vein

leaf blade

leaf margin entire, not lobed or divided

midvein

stem with node

petiole **Leaf parts**

scaly leaf

A rhizome has scaly, non-photosynthetic leaves

leaf margin lobed

## leaf cuttings

A number of herbaceous or woody plants can be propagated from a leaf.

Some leaves with a petiole, like those of African violets, can be propagated in water or in soil.

Rex begonias are multiplied by split vein propagation. A leaf is trimmed, and the main veins are cut. New plantlets with roots grow where the veins have been cut.

Monocotyledons have a series of parallel veins running the length of the leaf. Some, like snake plants and *Lachenalia*, can be propagated by cutting a leaf into segments. These are then placed in a suitable potting mix or water. African violet (*Saintpaulia*), a eudicot, can also be propagated this way. The part of the cutting towards the leaf base is where bulbils or roots will form.

WAYS OF PROPAGATING WITH LEAF CUTTINGS

African violet (*Saintpaulia*)

Place the leaf petiole in potting mix or water

watermelon peperomia (*Peperomia argyreia*)

Place leaf halves cut side down into potting mix. Plantlets with roots form along the cut edges.

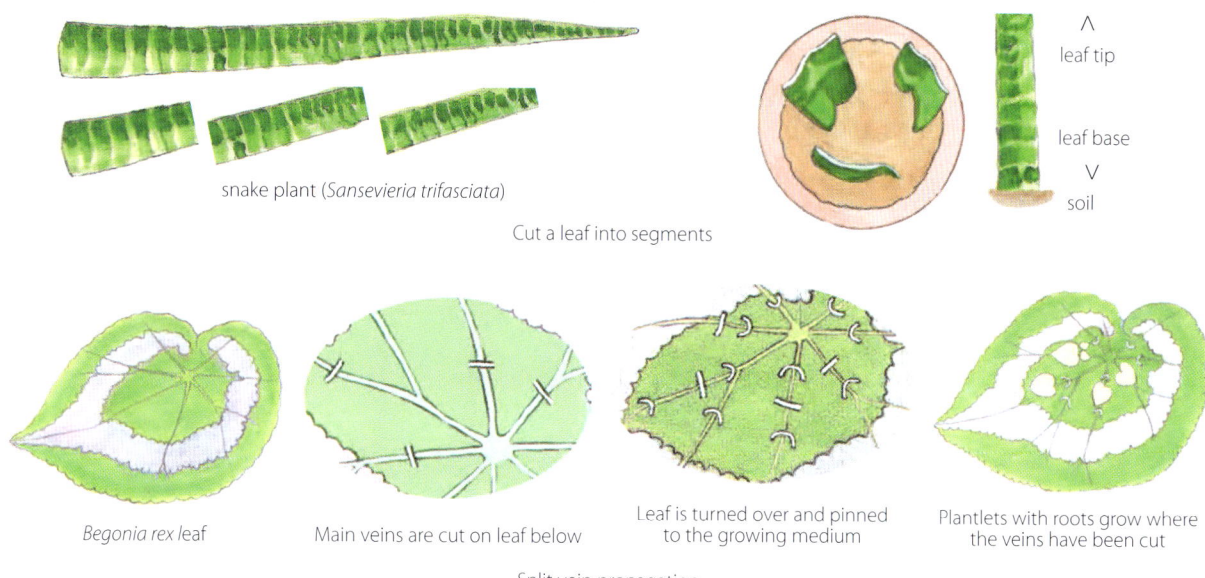

snake plant (*Sansevieria trifasciata*)

Cut a leaf into segments

^ leaf tip

leaf base

V

soil

*Begonia rex* leaf

Main veins are cut on leaf below

Leaf is turned over and pinned to the growing medium

Plantlets with roots grow where the veins have been cut

Split vein propagation

## leaf mould

Leaf mould is compost that is entirely made up of leaves.

It has a crumbly brown texture and a pleasant earthy scent. Leaf mould increases soil moisture as it can hold up to five times its own weight in water.

## leaf-bud cutting

A leaf bud cutting consists of a leaf, a petiole and a small piece of stem with a bud.

The cutting is placed in a growing medium, with the bud covered and the leaf exposed. The plant will grow from the bud. Rubber plants, clematis and rhododendrons can be propagated this way.

leaf

bud

petiole

stem

leaf-bud cutting

Cutting in the medium with the leaf only exposed

## leafminers *see* pests and diseases of gardens

## legginess *see* etiolation

## legumes

Legumes are plants that produce pods, the characteristic fruit of the pea family (Fabaceae).

They include peas, beans, soybeans, peanuts, wattles and ornamentals like wisteria. Alfalfa is grown as a cover crop, and pigeon peas have multiple uses including as animal fodder and windbreaks.

Legumes form a beneficial relationship with soil bacteria that fix nitrogen from the atmosphere and make it available to the plant.

*see also* nitrogen fixation

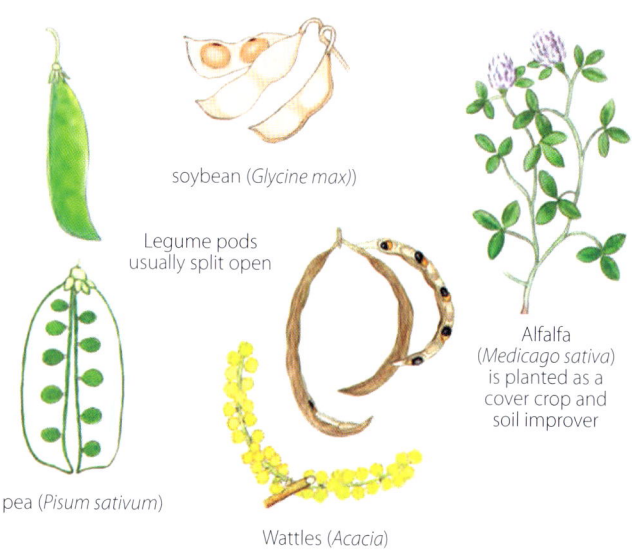

soybean (*Glycine max*)

Legume pods usually split open

pea (*Pisum sativum*)

Wattles (*Acacia*)

Alfalfa (*Medicago sativa*) is planted as a cover crop and soil improver

## life cycle

Most garden plants start life as a seed that germinates in the soil. The first leaves appear above ground, and the seedling grows into a mature plant with flowers. Flowers are pollinated and develop fruit with seeds. The plant survives by repeating the life cycle.

An annual plant completes its life cycle within a year then dies. Corn, cucumber and lettuce are examples.

A biennial plant completes its life cycle in two years, then dies. It will grow roots, stems and leaves in the first year and flowers, seeds and dies in the second year. Brussels sprouts, cabbages, carrots and parsley are examples. If left to grow for a second year, they will produce flowers and seeds before they die.

A perennial plant has a life cycle of more than two years. They can flower and fruit year after year. Apples, bananas and blueberries are examples. Many herbaceous perennials, like daffodils and jonquils, die down in winter and send up new growth in spring.

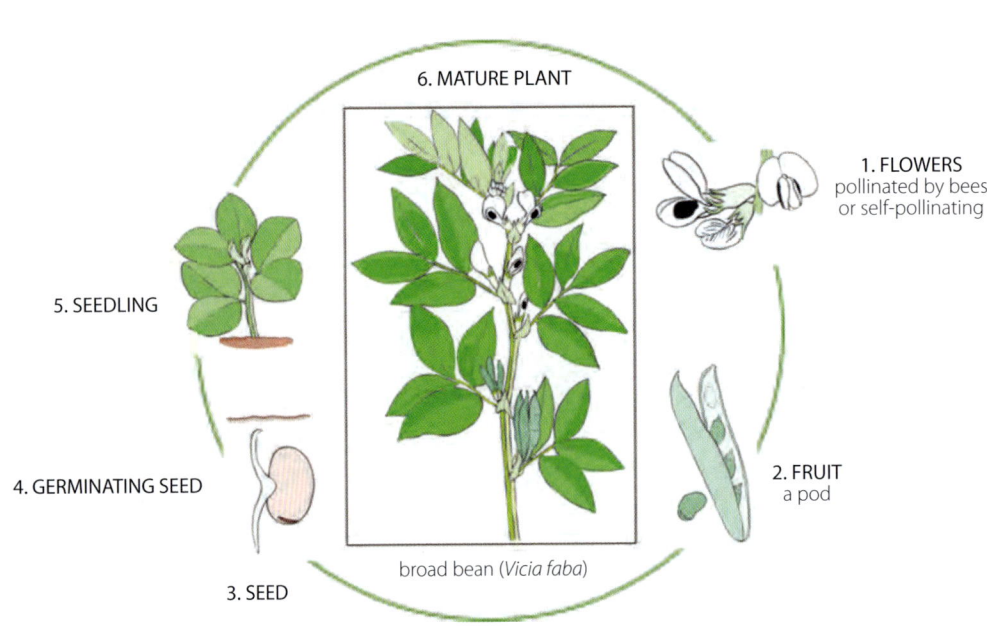

LIFE CYCLE

6. MATURE PLANT

1. FLOWERS
pollinated by bees
or self-pollinating

5. SEEDLING

2. FRUIT
a pod

4. GERMINATING SEED

broad bean (*Vicia faba*)

3. SEED

## lime

Garden lime is made from powdered limestone and is used to correct the pH of soils that are too acidic. Other forms of lime include calcitic lime, that contains calcium carbonate, and dolomite lime, that contains calcium carbonate and large amounts of magnesium.

**limey soil** *see* alkaline soil

**litmus paper** *see* pH, soil pH

**live stake cuttings** = **pole cuttings**

## loam

Loam is a mixture of clay, sand and silt. It is fertile, well-drained and easy to work. The ideal mix is mostly considered to be about equal parts sand, silt and clay. The ideal topsoil for gardening is loam.

**macronutrients** *see* plant nutrients

**magnesium** *see* plant nutrients

## male and female flowers

Male flowers have stamens that produce pollen, and have either no pistil or an undeveloped pistil.

Female flowers have one or more pistils that contain the ovary with ovules and have either no stamens or undeveloped stamens.

Male flowers and female flowers may be borne on different plants (a dioecious plant), like asparagus, holly and dates. Or separate male and female flowers may be on the same plant (a monoecious plant), as with cucumbers, melons and pumpkins.

Flowers that have only one functioning sexual structure are said to be imperfect.

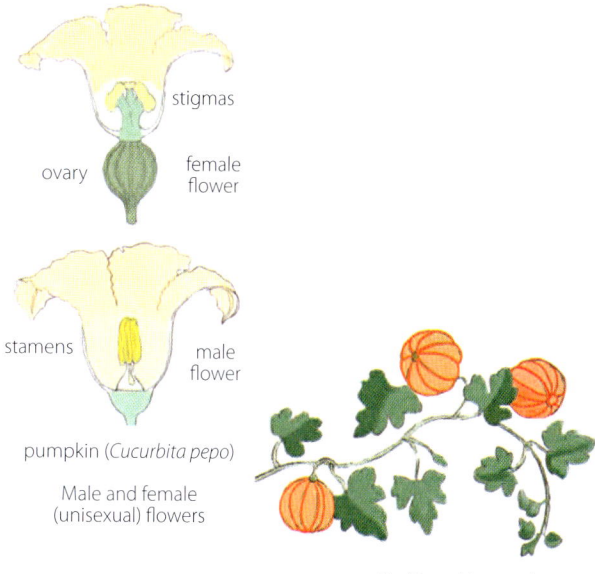

pumpkin (*Cucurbita pepo*)
Male and female (unisexual) flowers

pumpkin (*Cucurbita pepo*)

## mallet cutting

A mallet cutting is a section of last year's hardwood growth (the mallet), with a side shoot of the current season's semi-hardwood attached to it.

**manganese** *see* plant nutrients

## manure

Manure is solid and liquid waste from animals that are herbivores (plant eaters), typically cows, sheep, horses and chickens. Cat and dog manure is unsuitable because it is likely to carry parasites.

For garden use, manure should be composted for at least 6 to 12 months, until it turns dark, crumbly and odourless. Fresh manure has ammonia in it that will burn plants and bacteria that can contaminate edible plants.

Animal manure is a source of slow-release nutrients. It supplies primary nutrients (nitrogen, phosphorus and potassium) and micronutrients for plant growth. It is organic matter that improves soil structure and increases the water-holding capacity of sandy soils.

**marcotting** = air layering

## market gardens

A market garden is a small farm, usually under one acre, used for small-scale production of vegetables, flowers and fruits as cash crops. There are also large-scale commercial market gardens.

Backyard market gardens are becoming popular in Australia. Gardeners focus on crops for sale to restaurants, small local grocery stores, and to sell at local markets. Lavender and dried flowers are sold for craft products and aromatherapy. Some gardeners have greenhouses, and others concentrate on hydroponics. Vertical gardens save space in a backyard garden.

Food producers usually specialise in a select group of high value, quick-growing crops. Intercropping and succession planting make for intensive land use and continuous supply during the growing season. Cut and come again crops, like rocket, basil, parsley and leaf lettuce, produce over a long period of time.

### SOME BACKYARD MARKFT GARDEN CROPS

Nasturtium leaves and flowers are edible

Tomatoes

Thyme

salad mixes
rocket
spring onions
edible flowers
fresh herbs
dried flowers and herbs
tomatoes
seedlings

Spinach grown using succession planting

Seedlings

### maze

A maze is a network of paths and hedges with just one true route to the exit.

Unlike a labyrinth, a maze is designed with branching paths and hedge barriers to make finding your way difficult. The hedges are high enough so that the person walking through cannot see the general layout.

hedge —

A simple plan for a maze

### mealy bugs *see* biological pest control, pests and diseases of gardens

### Medieval gardens *see* page 85

### medium

In horticulture, a medium is an artificial soilless mix for growing plants.

The type of growing mix varies with its use, as for seeds and seedlings, cuttings, pots and hydroponics. Other plants in pots, like cacti, succulents and orchids, all have their own special medium requirements.

Many commonly used materials, like peat, perlite, vermiculite and rockwool, are now considered to be environmentally damaging. Sources of organic materials for media, like coir, a waste product of the coconut industry, composted organic municipal wastes that would otherwise go to landfill, mushroom compost, aged soft-wood pine bark, and sand are sustainable options.

### melons

Melons, also called pepos, are the fruit of the gourd family, Cucurbitaceae.

They include cucumbers, squashes and pumpkins, loofahs and watermelons. Most plants are annual vines with tendrils.

Flowers are showy and typically unisexual. Gardeners often cross-pollinate them by hand.

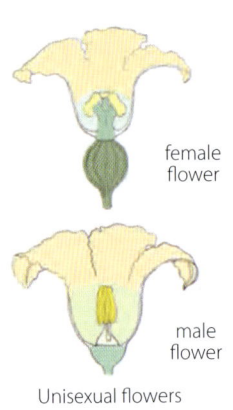

female flower

male flower

Unisexual flowers

cantaloupe

pumpkin

Trailing vine

## medieval gardens

The medieval monastery garden began in Western Europe with Charlemagne in c. 800 AD. It flourished, together with the feudal system of land management, between the 9th and 15th centuries.

In Europe, land was owned by the state (the king) and the church (bishops). The church followed the same feudal system as the state in allotting land (estates called fiefs) to princes and lords who lived in castles and manors. Some land was given to abbots to establish monasteries. Peasants worked the estates and had some land for their own use. Princes, lords and peasants paid a fee (fief) to the master of their land.

Gardens were laid out in square or rectangular plots with narrow paths between them. Typically, they were enclosed by stone walls, hedges, or pole fences. Kitchen gardens were sown with culinary herbs and many kinds of vegetables, including cabbage, melons, onions, spinach, eggplant, peas, lentils and beans.

Gardens were mainly for producing food. However, monasteries also had a physic garden with herbs (simples) for medicinal uses. The main buildings of a monastery were connected by a covered arcade (cloister) around an open courtyard garden (garth). Pleasure gardens for the rich and secluded gardens in monasteries had shade trees and lawned spaces planted with decorative plants like roses, violets and sweet-smelling herbs.

Fruit trees were espaliered along walls, interlaced to form a hedge or cover-way, or planted in orchards in geometric patterns. Pietro de Crescenzi (c. 1235– c. 1320) was one of the first to formally write about horticulture. He advised that in orchards *'Trees are to be planted in their rows, pears, apples, & palms, & in warm places lemons. Again mulberries, cherries, plums, & such noble trees as figs, nuts, almonds, quinces & such-like, each according to their kinds, but spaced twenty feet apart more or less.'* Trees like beech were coppiced to provide poles for fences, and other deciduous trees were pollarded for poles, wood and craft supplies.

## MEDIEVAL GARDENS

Gardens with square or rectangular plots

carrot

sweet melon

aubergine
Vegetable garden

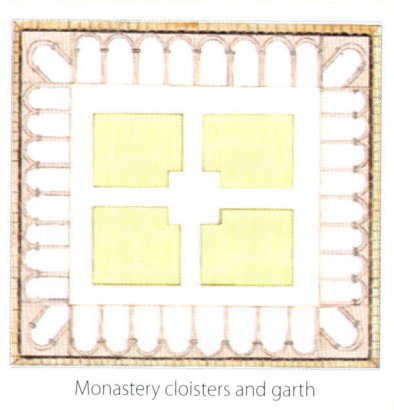

Monastery cloisters and garth

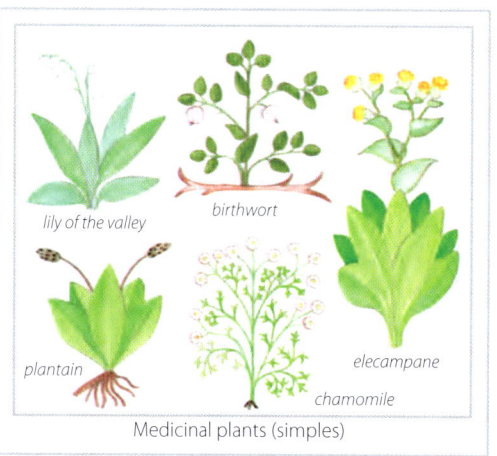

lily of the valley

birthwort

plantain

chamomile

elecampane

Medicinal plants (simples)

Pleasure garden

## microclimate

A microclimate is an area in a garden with a different climate to its surroundings. It could be shady or sunny, sheltered or exposed to wind.

## microgreens

Microgreens are the cotyledons and first true leaves that appear above ground after a plant has germinated.

The cotyledons and leaves are cut off at soil level and washed before being eaten. Seeds grown for microgreens include watercress, alfalfa and radish.

## micro-irrigation *see* irrigation

## micronutrients *see* plant nutrients

## micropropagation

Micropropagation is a way of reproducing plants from cells in a small bud, shoot, or other piece of tissue.

The piece to be propagated is sterilised and placed on a nutrient medium in a glass or plastic container. When new plantlets have roots and leaves, they are ready to be grown on in tubes and placed in a propagator. Individuals are later planted out separately and hardened off, ready for planting outside.

Micropropagation is usually carried out in a laboratory, but some plants can be propagated this way by a gardener using special equipment.

## misting

Misting is done with nozzles attached to a water pipe. They make a fine mist to lower greenhouse temperatures and maintain humidity for plants.

It is an ideal system for warm, dry climates but is unsuited to climates with a naturally high relative humidity and high temperatures.

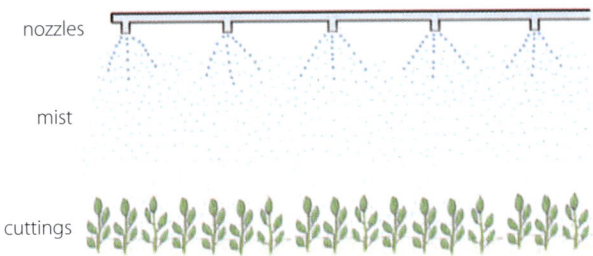

## molybdenum *see* plant nutrients

## monocotyledons *see* page 87

## monoculture

Growing a single species of plant over a wide area for mass-produced food. A field of wheat, sugarcane or carrots is a monoculture.

## monoecious

Having separate male and female flowers on the same plant, like cucumbers, melons and pumpkins.

## monocotyledons

Monocotyledons are the second largest group of flowering plants (22% of all angiosperms).

They are characterised by a seed with only one cotyledon (seed leaf). Flower parts are usually in multiples of three, with the petals and sepals usually similar. Leaves have parallel main veins. Roots are usually fibrous.

*cf.* **dicotyledons, eudicots**

MONOCOTYLEDONS

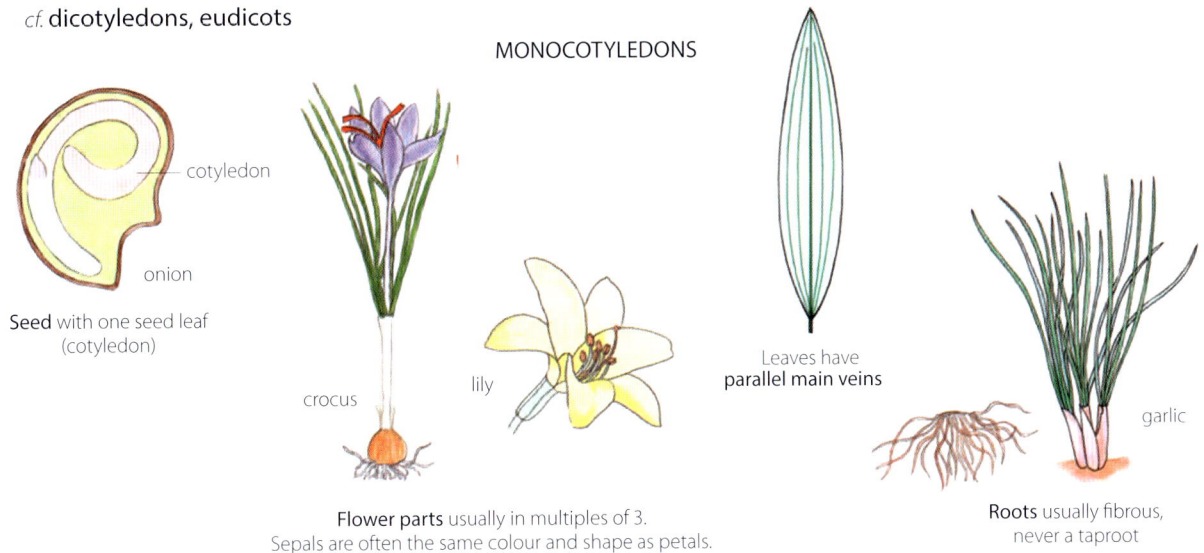

cotyledon

onion

**Seed** with one seed leaf
(cotyledon)

crocus

lily

**Leaves** have
**parallel main veins**

garlic

**Roots** usually fibrous,
never a taproot

**Flower parts** usually in multiples of 3.
Sepals are often the same colour and shape as petals.

## moon gardening

Moon gardening is the practice of planting, harvesting, cultivating and fertilising in harmony with the moon. The phases of the moon determine the best time to do these tasks.

Just as the moon influences the rise and fall of the tides, its gravity also affects the moisture in plants and in the soil. These effects vary at different times in the moon cycle. For example, the best time for planting leafy annuals, like basil and rocket, is during the new moon phase when water is being drawn upwards.

The idea that the moon affects plant growth is an ancient one. Moon calendars are available today, some with great detail according to star signs and tasks to be done on particular days.

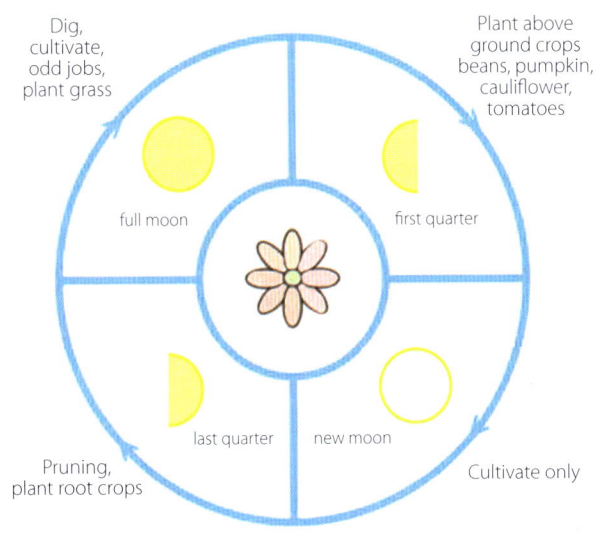

Dig, cultivate, odd jobs, plant grass

Plant above ground crops beans, pumpkin, cauliflower, tomatoes

full moon

first quarter

last quarter

new moon

Pruning, plant root crops

Cultivate only

Moon gardening calendar

## mound layering

In mound layering, also called stool layering, a plant is cut back to nearly ground level. As the new shoots develop, they are covered with layers of soil to encourage root development.

Rooted shoots are later separated and grown as new plants.

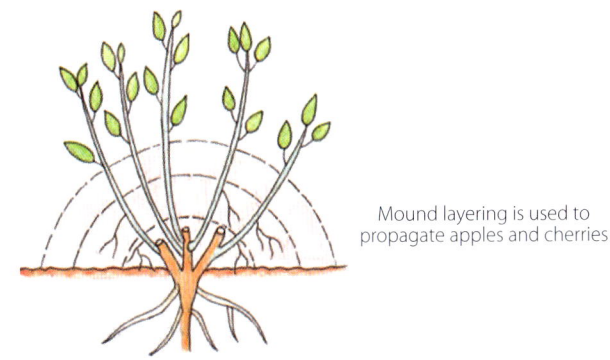

Mound layering is used to propagate apples and cherries

## mulch

Mulch is a covering of organic material (compost, rotted manure, shredded bark, shredded sugar cane) or non-organic material (gravel, polythene sheeting, old carpet) placed on the surface of cultivated soil.

It suppresses weeds and conserves water by reducing evaporation. Organic mulches break down over time and improve the soil.

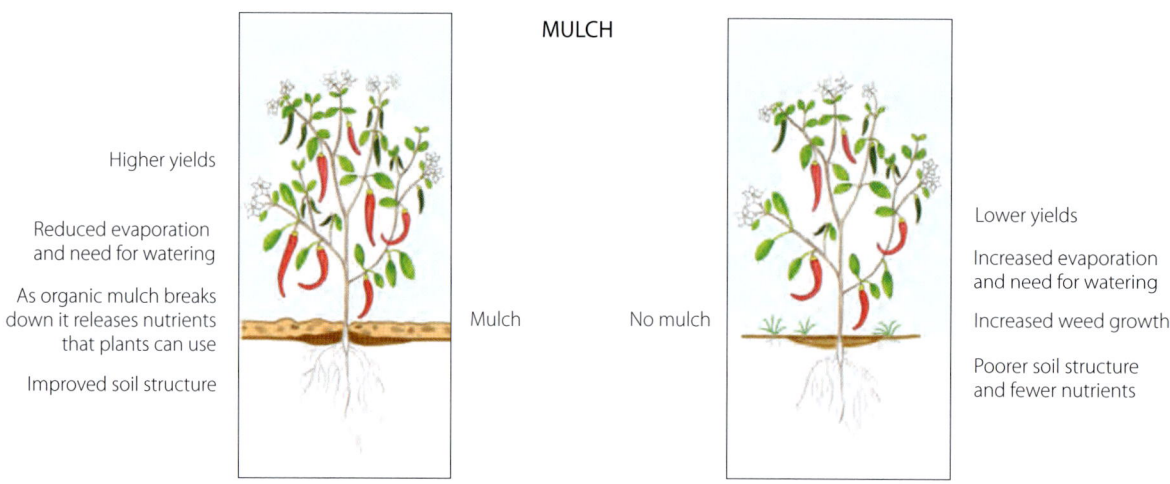

MULCH

Higher yields

Reduced evaporation and need for watering

As organic mulch breaks down it releases nutrients that plants can use

Improved soil structure

Mulch

No mulch

Lower yields

Increased evaporation and need for watering

Increased weed growth

Poorer soil structure and fewer nutrients

## mushroom compost

Mushroom compost is an organic by-product of mushroom farming.

The compost might include decomposed stable bedding and manure, wheat straw, poultry litter and cotton seed hulls. The mix is pasteurised to kill any harmful bacteria. It releases nutrients slowly and improves soil structure, as well as conserving soil moisture.

## mutation

A mutation is a change in the genetic makeup of a plant that can alter its appearance or other qualities like disease resistance. It can occur spontaneously in nature or be induced artificially by radiation and chemicals. Many new varieties of fruits and vegetables are developed artificially by being genetically modified in this way.

## mycorrhiza

A mycorrhiza is a partnership between the thread-like hyphae of a fungus and the roots of a plant.

The threads of hyphae enter the roots of the plant and supply water and mineral nutrients, like phosphorus. In turn, the hyphae receive nutrients from the plant, like carbon that results from photosynthesis.

There is a vast network of hyphae between plants that shares nutrients in the soil. Hyphae also store water that plants use in drier times.

From time to time, the underground hyphae send up a fruiting body, like a mushroom, that produces reproductive spores. Some, like truffles, have their fruiting bodies underground. The truffle fungus has a mycorrhizal association with the roots of oak trees.

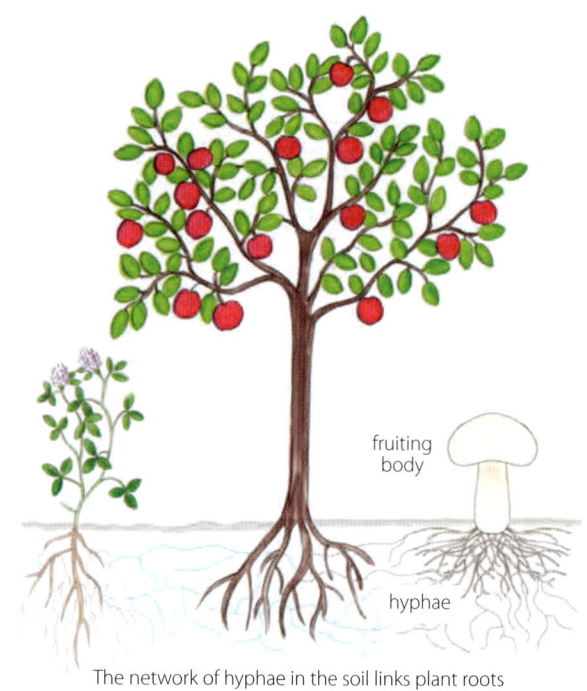

fruiting body

hyphae

The network of hyphae in the soil links plant roots

### native gardens

Native gardens use plants that grow naturally in a particular place and are not introduced from elsewhere, often another country. In Australia, they often feature local bush tucker foods.

### natural pest control

Natural pest control methods rely on remedies that use naturally occurring organic substances and physical controls like fences, traps, and horticultural fleece.

Organic controls include companion plants that can attract beneficial insects, repel pests, provide nutrients, and give physical support or shade to other plants. There are theories that plants can chemically enhance or inhibit each other's growth and well-being. Natural predators can be introduced into the garden to reduce the numbers of a particular pest. There are also natural pesticides, like neem oil, pyrethrum and diatomaceous earth.

Substances can have negative side effects and be harmful despite being organic. Check the instructions carefully before using them.

*see also* biological pest control, companion planting

### naturalised plants

Naturalised plants are exotic plants that can grow on their own in a new environment. They may be considered good, like swathes of daffodils that come up in spring, or bad, like blackberries that are invasive.

### necrosis

Necrosis is localised death of cells that leaves the surrounding plant tissue unaffected. It is a symptom of fungal infection and nutrient deficiency.

Leaf with necrosis
*Malus domestica* 'Golden delicious'

### nectar

A sugary substance, secreted by glands (nectaries), that attracts pollinators.

### nectary

A nectary is a gland that secretes nectar.

Nectaries are located on flowers, like those at the base of the petals of buttercups (*Ranunculus*).

Some are located on plant parts other than the flower, such as those on the petiole of cherries (*Prunus avium*).

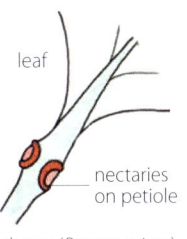

leaf

nectaries on petiole

cherry (*Prunus avium*)

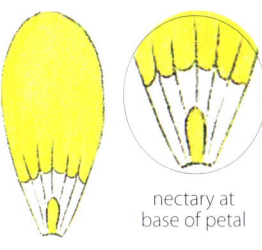

nectary at base of petal

buttercup (*Ranunculus*)

### neem oil

Neem oil is pressed from the seeds of the neem tree (*Azadirachta indica*). It has been used for centuries as a safe and powerful pest control method.

It is available as a foliar spray or a soil drench. The compound azadirachtin is absorbed into leaf tissue and repels chewing and sucking insects. It does not target insects like butterflies and ladybirds that do not chew on leaves. There has been some concern about the use of neem oil and its effects on bees. Instructions for use should be followed carefully.

## nematodes

Thousands of nematode species live in the soil. They are minute round worms, sometimes called eelworms. They feed on fungi, bacteria and other soil organisms, and some attack and kill pests like grubs, thrips and beetles.

Most are beneficial, but some are harmful. Pest nematodes, like the potato cyst nematode and the root-knot nematode, cause disease by infesting the plant's roots.

nematodes

Nutrients like nitrogen and phosphorus are stored in the bodies of bacteria, fungi and other organisms. When nematodes eat and digest them, these nutrients are released into the soil in a form that can be used directly by plants.

## neutral soil

A neutral soil is neither acidic nor alkaline and has a pH between 6.0 and 7.5. Most plants thrive best in a neutral soil.

## nickel *see* soil nutrients

## nitrogen, (N), (N$_2$)

Nitrogen is a gas that makes up 78% of the Earth's atmosphere. Nitrogen diffuses into the soil from the atmosphere and needs to be changed or 'fixed' before plants can use it. It is then absorbed through their roots in the form of nitrate and ammonium.

## nitrogen fixation

Nitrogen diffuses into the soil from the atmosphere and needs to be changed or 'fixed' before plants can use it. In this process, bacteria link nitrogen to oxygen to form nitrate and nitrogen to hydrogen to form ammonium, both of which can be used by plants.

Nitrogen-fixing organisms, like the bacteria *Rhizobium* in the root nodules of legumes such as clover, fix nitrogen for the plant's own use.

Lightning fixes a small amount of nitrogen that enters the soil with rainfall.

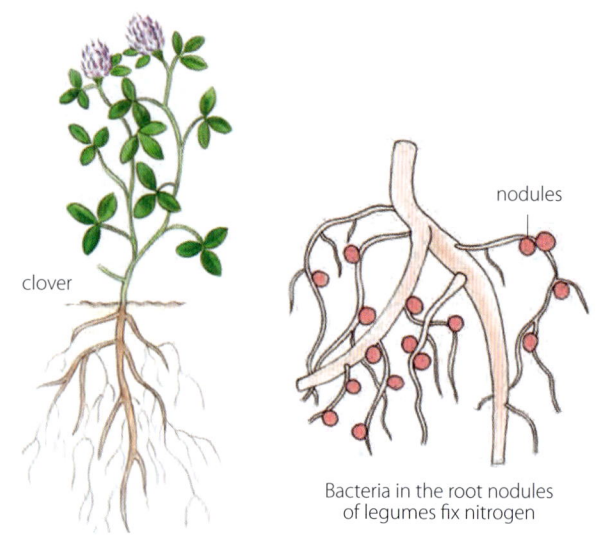

clover

nodules

Bacteria in the root nodules of legumes fix nitrogen

## no dig gardens *see* page 91

## node

All stems, whether above- or below-ground, have nodes from which leaves, shoots and aerial stems grow.

The space between two nodes is the internode.

*see also* **stem**

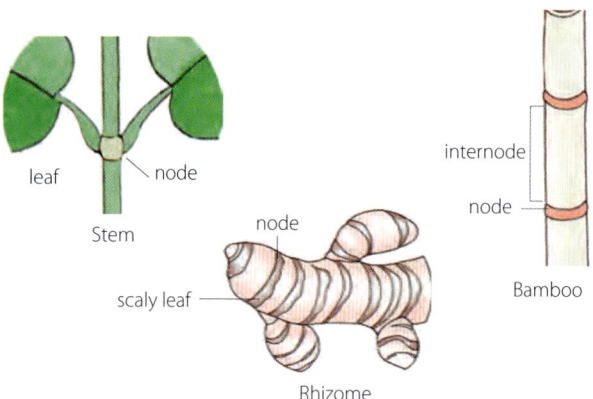

leaf

node

Stem

node

scaly leaf

Rhizome

internode

node

Bamboo

## no dig gardens

No dig gardens were developed in Australia by Esther Dean in the 1970s. Since 1981, Charles Dowding, a leading innovative gardener in south-west England, has used his garden at Homeacres to promote organic, no dig gardening worldwide. Digging soil disturbs its structure and its organisms. Instead of regularly turning over the soil, a no dig garden leaves the soil undisturbed and replenishes it by layering compost on the surface.

Soil is a complex ecosystem with an abundant and diverse life. With no dig gardening, this ecosystem remains undisturbed. Worms can continue to aerate the soil and provide good drainage. Microorganisms and the complex network of fungi, that extends over large areas and connects all soil life, are unharmed. There are fewer weeds as seeds remain in the soil instead of being brought to the surface when digging.

A no dig garden can be constructed anywhere: on a lawn, over an existing bed, or on clay or other poor soil. It is more productive per square metre, plants are healthier, and it is less labour-intensive for the gardener.

### MAKING A NO DIG GARDEN BED

Different organic layers are laid down and rot over time, usually about 6 months, to form a rich soil that is ready to be planted

**An example of layers**

| | |
|---|---|
| Layer 5 | Mulch |
| Layer 4 | **Compost** layer |
| Layer 3 | **Brown layer** of organic matter like leaves, coir and straw, and wet thoroughly |
| Layer 2 | **Green layer** of nitrogen-rich matter like grass clippings and manure, and wet thoroughly |
| Layer 1 | **Cardboard and newspaper**, well-watered, to suppress weeds |

The green and brown layers can be repeated to make up the required depth

## nodule

A small round swelling on the roots of legumes. They are formed by nitrogen-fixing bacteria in the root.

*see also* **nitrogen fixation**

## NPK ratio

Of fertilisers, the percentage by weight of nitrogen (N), phosphorus (P) and potassium (K). Nitrogen for photosynthesis, phosphorus for root growth, nutrient and water transport and potassium for growth and reproduction.

## offset

Offsets are daughter plants that grow from the base of the parent plant.

They are clones of the parent, having the same characteristics, and can be detached for propagation.

Offsets are commonly called 'pups'.

Banana offsets are suckers

Globe cactus with 'pups'

Gladiolus with cormel offsets

## open pollination

Open pollination refers to plants that are pollinated naturally, as by wind, birds, and bees or other insects. Open-pollinated seeds have the same characteristics as their parents and are said to be 'true to type'.

## orchids

Orchids make up the largest family of flowering plants on Earth. They are intriguing because of their complexity, unique beauty and diversity.

Most orchids are native to the tropics, where they attach themselves to the bark of trees or to other plants (epiphytic orchids). Other orchids grow in the ground (terrestrial orchids) or on rocks (lithophytic orchids).

The moth orchid (*Phalaenopsis*) is one of the world's most popular pot plants. Like all orchids, it has three petals and three sepals, and a column that combines the male anthers and the female stigma. One of the three petals, the labellum, is different from the other two and is highly modified to attract pollinators. Dust-like seeds develop in the ovary below the flower.

Many orchids are difficult to grow, especially some native Australian terrestrials that rely on a specific pollinator and have a specific relationship with fungi in the soil.

However, numerous beautiful hybrids and cultivars have been developed, and there are many species that can be grown successfully. It is essential to study their individual requirements for soil, temperature and where they are happy to grow.

There are six main ways to propagate orchids: division, back bulbs (as *Cymbidium*), keikis (as *Phalaenopsis* and *Dendrobium*), aerial cuttings, tissue culture and seed. Both tissue culture and seed propagation require sterile conditions and are best done in a laboratory. Some orchids, like the blunt greenhood (*Pterostylis curta*), form colonies from underground daughter tubers. Plantlets can be separated and potted up to form new colonies.

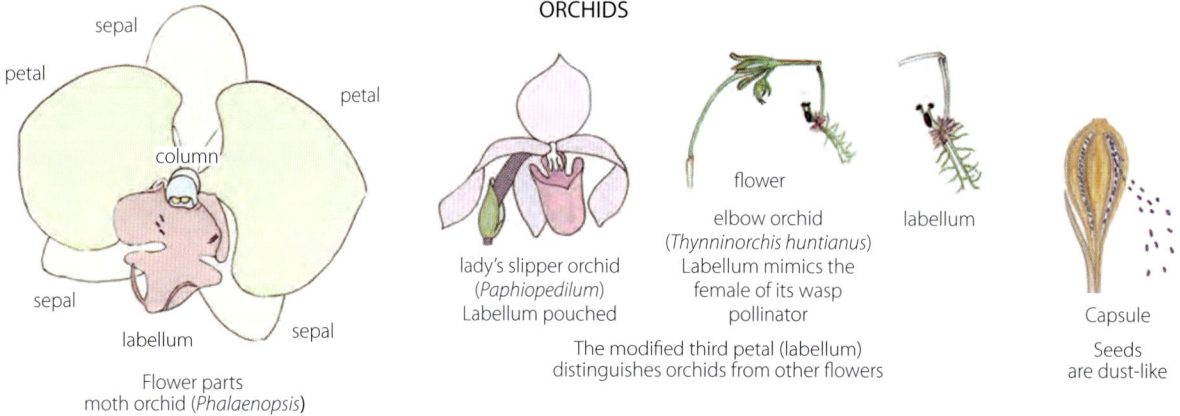

ORCHIDS

sepal
petal
petal
column
sepal
labellum
sepal
Flower parts
moth orchid (*Phalaenopsis*)

lady's slipper orchid
(*Paphiopedilum*)
Labellum pouched

flower
elbow orchid
(*Thynninorchis huntianus*)
Labellum mimics the
female of its wasp
pollinator

labellum

The modified third petal (labellum)
distinguishes orchids from other flowers

Capsule
Seeds
are dust-like

## organic fertiliser *see* fertiliser

## organic gardens

Organic gardens rely on resources that were once living plants or animals. Artificial fertilisers that feed the plant but not the soil and commercial solutions using chemicals to control pests, diseases and weeds are not used.

Building good soil is the key to a fruitful organic garden. Compost mulches are a rapid source of nutrients for soil organisms like worms and microbiota. They break down and give good texture and structure to the soil, allowing it to hold the water and air that plants need. Threads of fungi spread through large areas of soil and provide roots with food and moisture. These, as well as the soil biota, work best if left undisturbed by digging.

Organic gardeners encourage biodiversity by growing a wide variety of plants. Plant-covered soil provides less space for weeds.

Other features of organic gardens might include companion planting that helps control pests, and raising animals like goats and chickens to provide manure. Cover crops like clover can be grown in winter when the soil would otherwise be bare. They provide green manure, and their nitrogen-rich roots die down and feed the soil. Catch crops, like lettuce, that grow quickly, will provide a harvest while other plants, like beans, are maturing.

Any kind of garden is suitable for organic gardening. It works for food plants, trees and ornamentals. It might be an elevated garden bed on a balcony, a domestic backyard garden, or an acreage.

### organic matter

Organic matter refers to anything that comes from living or once-living plants and animals.

### organic matter in soil

Soil organic matter consists of small fresh plant and animal residues, living soil organisms, actively decomposing organic matter, and stable organic matter (humus).

It makes up about 5–10% of soil suitable for growing plants, the other components being minerals, water and air.

Soil organic matter is a reservoir of nutrients for plants. It aids soil aggregation and aeration and retains moisture. It also increases water infiltration into the soil.

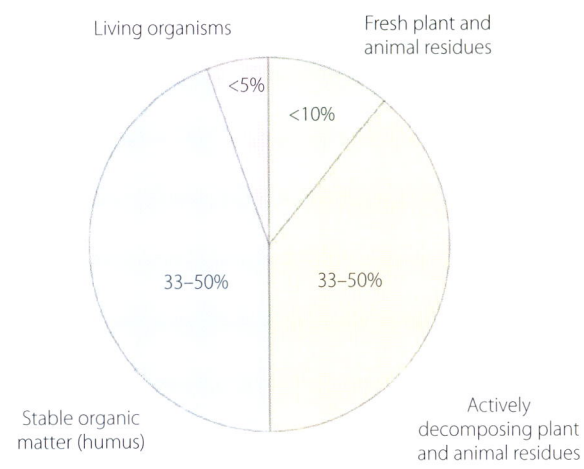

Living organisms  <5%
Fresh plant and animal residues  <10%
33–50%
33–50%
Stable organic matter (humus)
Actively decomposing plant and animal residues

### organic pest control *see* natural pest control

### ornamental gardens

Plants in ornamental gardens are chosen for the beauty of their foliage and flowers rather than for food.

The layout varies. The gardens at the Palace of Versailles in France are extremely formal and elegant. The layout is geometrical and favours symmetry. Shrubs are topiaried and hedges are trimmed.

The cottage garden is an informal ornamental garden of colourful flowers and flowering trees. Plants are allowed to grow into their natural shape. Rose gardens are ornamental gardens, as are perennial borders and grassy expanses with naturalised bulbs.

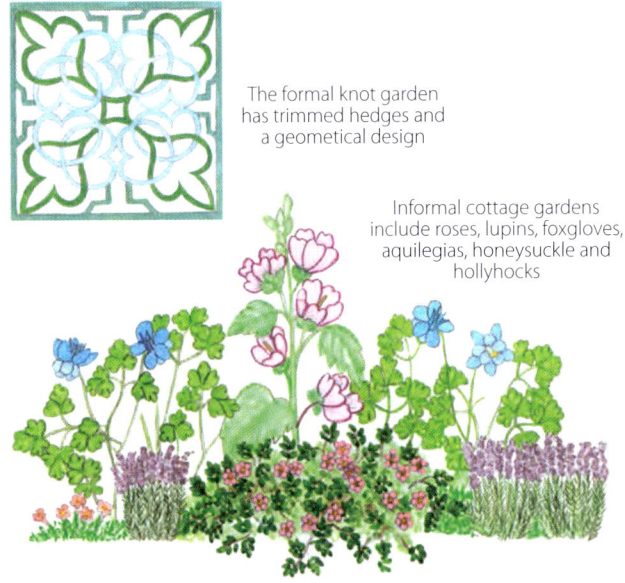

The formal knot garden has trimmed hedges and a geometical design

Informal cottage gardens include roses, lupins, foxgloves, aquilegias, honeysuckle and hollyhocks

### outer bark

Outer bark is the dead tissue on the outer surface of a woody plant.

**ovary** *see* flower

## over-seeding

Spreading grass seeds over the top of an already established lawn to thicken it up and remedy sparse areas. It avoids having to plant a completely new lawn.

## ovule

An ovule is an immature seed that develops fully after fertilisation.

It is located in the ovary of flowering plants (angiosperms) and on the cone scale of non-flowering seed plants (gymnosperms).

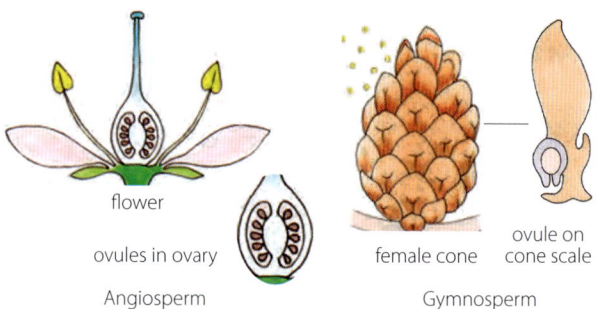

flower

ovules in ovary

Angiosperm

female cone

ovule on cone scale

Gymnosperm

## palms

The common name for a member of the Arecaceae family. Palms can be tree-like, shrubs or climbing vines (rattan). They are mostly native to tropical and subtropical regions.

Palms have a pithy single stem or multiple stems with leaf scars, formed as fronds fall from the growing plant. Except for rattans, palms have a crown of divided fronds. They have an inflorescence of small bisexual or unisexual flowers, and the fruit is either a drupe (dates) or berry-like (acai berry). Roots are fibrous and close to the surface.

Date palms and coconut palms are widely grown for food. The acai palm is cultivated for its berry-like fruits and hearts of palm. The heart is the edible growing tip of some palms, and if harvested from a single-stemmed plant, the palm will die.

The acai palm is multi-stemmed and selecting a few stems for harvest and leaving others will not kill the plant.

Palms are popular ornamental plants. *Chamaedorea* is a large genus that is native to the tropical and subtropical regions of Central and South America. Most are smaller in size, and a few species are among the most commonly grown indoor and outdoor ornamental palms.

crown of divided fronds

inflorescence of small flowers

feather-like pinnate frond

stem pithy

fan-like palmate frond

stem with leaf scars

fruit berry-like acai berry

roots fibrous

fruit a drupe date

## parent material

Parent material is the original material from which the soil profile develops. It may be bedrock or unconsolidated material. Loose materials like river alluvium, wind-blown silt particles, volcanic ash and organic matter may be deposited on this, but derive from different parent material.

## parterre

The parterre was first developed in late 16th century France by garden designer Claude Mollet. He borrowed the geometric designs of the Italian Renaissance plots and added complex patterns inspired by embroidery, to create the 'parterre de broderie'. It became a central feature of formal gardens in France in the 17th century and was widely copied. In England, it became an evolution of the knot garden.

The patterns were the major feature and were most impressive when viewed from above. Hedged beds were laid out on level ground and separated by gravel paths. They enclosed colourful annuals, perennials and herbs. In winter the greenery of the hedges remained a pleasing feature.

The decorative parterre designs are used widely in contemporary gardens.

PARTERRE

C17th design by André Le Nôtre showing parterres of the Tuileries Gardens in Paris

Detail of a 'parterre de broderie'

## patch budding

Patch budding is used on plants, like pecans, with thick bark that can be easily separated from the wood.

It is done on small tree trunks or branches 1–4 cm in diameter. Larger trees that have been cut back, like olive trees, can be patch budded with vigorous new shoots.

A small, rectangular patch of bark with a single bud (scion) is taken from the plant to be propagated. It is then placed into the space where a similar patch has been removed on the rootstock. The cambium of each is matched, and the two are then taped together firmly.

cambium

A rectangular patch is removed from the rootstock

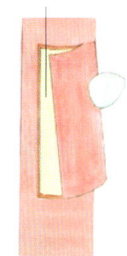
cambium

A scion is taken from the plant to be propagated

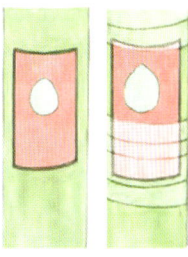

The scion is inserted in the rootstock and secured with tape

## peat moss

Peat moss is peat that is used in potting mixes and garden soils. It is acidic and increases the acidity of the soil, making it ideal for growing acid-loving plants like blueberries.

Peat moss begins life as sphagnum moss and other plant matter that dies to form a compacted mass over many hundreds or even thousands of years. The sphagnum moss wetland survives on top of it, and when the plants die, they contribute about 1–2 mm each year to the mass below.

Peat is saturated with water. Water that is no longer retained after peat moss is removed can cause flooding.

One third of the Earth's soil carbon, fixed from the atmosphere into plant tissue through photosynthesis, is locked away in peat.

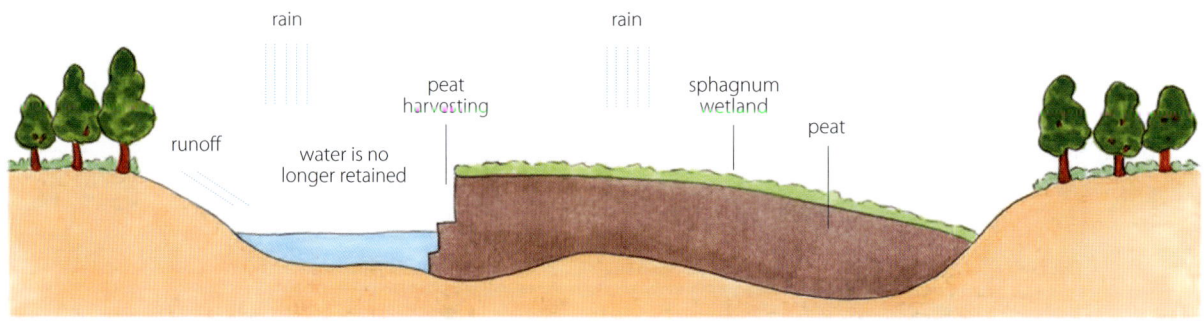

**peds** *see* soil aggregates

**pedicel** *see* inflorescence

**peduncle** *see* inflorescence

**pepo** *see* melon

**perennial** *see* life cycle

**perfect flower**
A perfect flower is a bisexual flower that has functioning stamens and pistils.

**peristyle**
A peristyle is a courtyard or inner garden, surrounded by a covered walkway supported by rows of columns. The space is open to the sky. The ancient Greeks and Romans built them in temples, public buildings and in houses.

**perlite**
Perlite is a hard, highly porous material made by superheating volcanic glass. It aerates the soil and holds moisture for a short time.

**permaculture** *see* page 97

**Persian gardens**
The Persian Empire lasted from about 559 BC to 331 BC and covered present day Iran, Egypt, Turkey, and parts of Afghanistan and Pakistan. Much of this land was desert, and walled gardens provided relief from summer heat and winter cold.

The oldest known Persian garden, built c 550 BC., is the palace garden of Cyrus the Great at Pasargadae in what is now Iran. It featured a symmetrical garden of four squares or rectangles. These were intersected by stone irrigation channels that transported water from a canal linked to a nearby river. It is considered to be the first chahar bagh, meaning four gardens. Later gardens channelled water from springs and qanats.

After Alexander the Great conquered Persia in 334 BC, the Greeks adopted the Persian garden design. More than a thousand years after Pasagardae, in the 6th century AD, the chahar bagh became associated with Islamic gardens and later, from 1526 to 1761, the Mughal Empire in India. Persian garden structures influenced the gardens of the Allambra in Muslim Spain and the Taj Mahal in Mughal India.

*see also* **chahar bagh garden, Islamic gardens**

96

## permaculture

Permaculture is an approach to gardening that aims to have animals, insects, humans, plants and microorganisms living in harmony in a healthy, self-sufficient environment.

The principles include:
- Imitating nature and minimising intervention from people.
- The concept of a food forest with layers of trees, shrubs, bushes and ground covers that regenerate naturally.
- Valuing diversity.
- Conserving the soil and water.
- Eliminating waste.

There is a strong foundation of design and planning that includes:
- Taking into account wind direction, sun, microclimates, rainfall and soil.
- Zones that are near or distant from the dwelling according to frequency of use.
- Intensive use of space with diverse plants mixed together so that they benefit each other. Permaculture guilds are an example of this.
- Using elements that have more than one function, like hedges of fruiting bushes, like cumquat, and flowering shrubs like rosemary that attract bees and are pleasing to look at.

Bill Mollison and David Holmgren coined the term 'permaculture' and developed the concept in the 1970s. They published their first book, *Permaculture 1*, in 1978. Geoff Lawton has expanded on their work. It has since been applied worldwide, in home gardens, on farms, and in reclaiming degraded land.

### PERMACULTURE ZONES FOR A HOME GARDEN

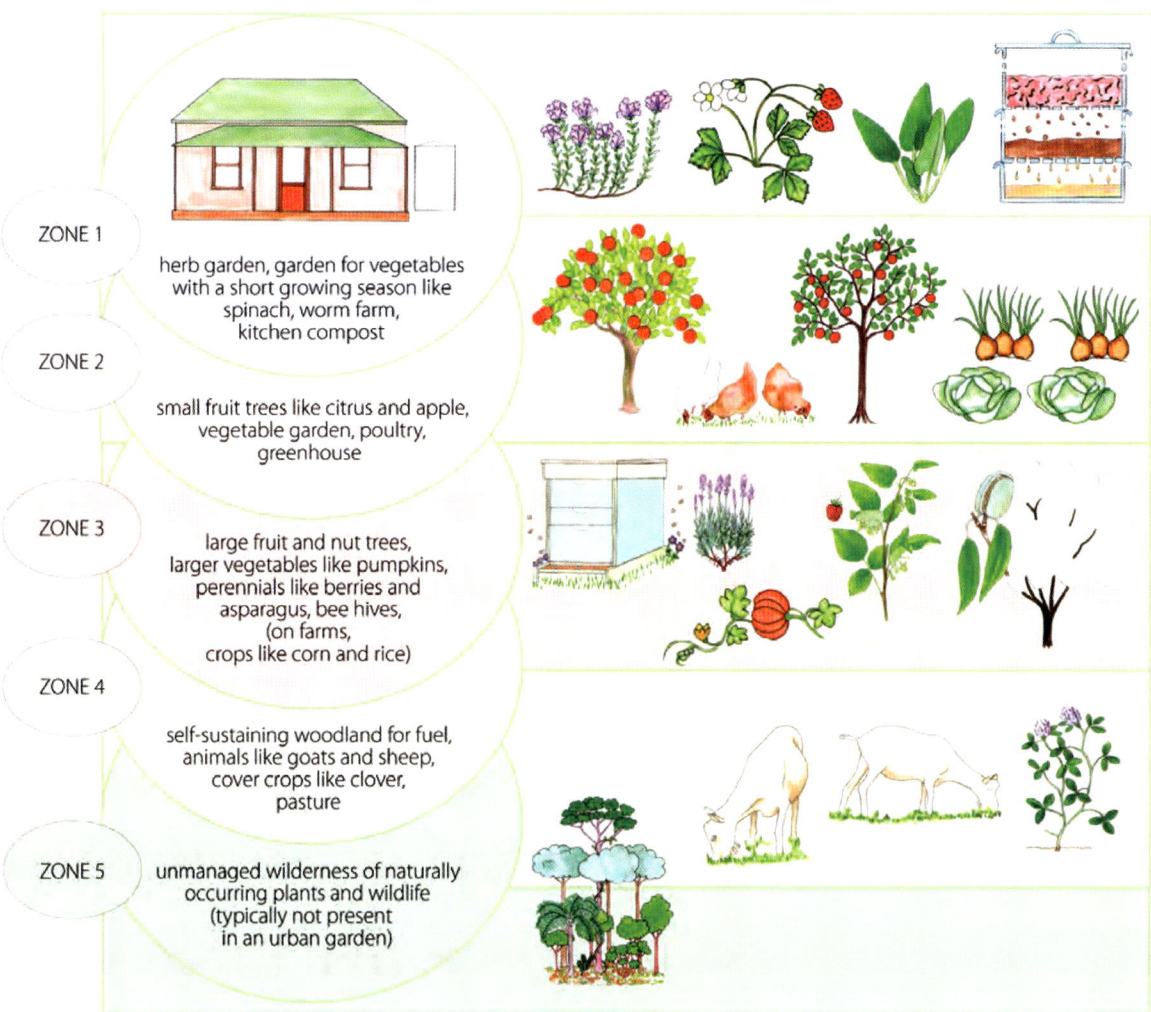

ZONE 1

herb garden, garden for vegetables with a short growing season like spinach, worm farm, kitchen compost

ZONE 2

small fruit trees like citrus and apple, vegetable garden, poultry, greenhouse

ZONE 3

large fruit and nut trees, larger vegetables like pumpkins, perennials like berries and asparagus, bee hives, (on farms, crops like corn and rice)

ZONE 4

self-sustaining woodland for fuel, animals like goats and sheep, cover crops like clover, pasture

ZONE 5

unmanaged wilderness of naturally occurring plants and wildlife (typically not present in an urban garden)

## pesticides

Pesticides are chemical substances used to destroy organisms that are harmful to cultivated plants. They are grouped according to the type of pest they kill, like insecticides for insects, herbicides for weeds, and fungicides for fungi.

Biodegradable pesticides can be broken down, by microbes and other living organisms, into harmless compounds. Others are persistent and can take months or years to break down. Some, like DDT, have been banned because they remain in the soil and water for years and can accumulate in the food chain.

All pesticides are potentially toxic and need to be used safely and disposed of properly.

*see also* **biological pest control, diotomaceous earth, natural pest control**

## pests and diseases of gardens

Pests mostly damage plants by how they feed.

Pests that feed by piercing and sucking include white flies, aphids and mealy bugs. Slugs and snails feed by licking or rasping away the surface of seedlings, soft leaves and young roots. Biting and chewing creatures, like grasshoppers and caterpillars, can cause dramatic plant damage. Leafminers lay eggs on the surface of leaves and bark. When the tiny larvae hatch, they tunnel below the surface and feed, making a characteristic scribbly pattern as they go. Cabbage white butterfly caterpillars eat leaves, and particularly like brassicas.

Fungi, bacteria and viruses cause other diseases. Wilt is caused by fungi and bacteria invading the water carrying vessels (xylem) of the host plant. The fungus genus *Fusarium* affects tomatoes, eggplants, peppers and potatoes. The fungus *Ophiostoma* causes wilt in elm trees (Dutch Elm Disease). The bacteria genus *Erwinia* causes wilt in the melon family, Cucurbitaceae. Viruses are usually transmitted from one plant to another by a carrier, usually a sucking insect. Aphids carry various mosaic viruses that infect beans, cucumbers and sweet potato. The greenhouse white fly transmits a virus that affects lettuce, beet, endive and cucumber.

Damage to plants can also result from deficiencies of plant nutrients in the soil, sunburn and over-fertilising. Lack of water will cause wilt.

*see also* **fruit fly, plant nutrients**

### SOME PESTS AND DISEASES OF GARDENS

SAP-SUCKING INSECTS

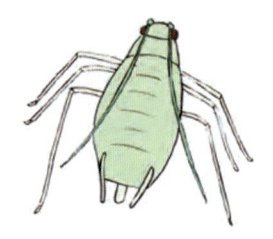

**Aphids**
are small, sap-sucking insects that infect new shoots, buds, leaves and stems. They secrete sugary honeydew that attracts ants and sooty mould. Some attack only specific plants, like the rose aphid and the cabbage aphid.
Aphids, mealy bugs and scales are the most common sap-sucking insects in the garden.

**Black spot**
isn't just a fungal disease of roses, it can also affect fruit trees, such as citrus, paw paw and apples. Spores travel on the wind and spread with splashes when watering.

**White fly**
are sap-sucking insects that multiply rapidly and lay their eggs on the underside of leaves. The baby nymphs and adult flies feed on leaf, bud and stem sap. They secrete honeydew that attracts ants and is food for sooty mould.

## FUNGAL DISEASES

**Powdery mildew**
is a common fungus that affects a wide variety of plants. It appears as white powdery spots mostly on leaves, but also on some stems, flowers, fruit (like grapes) and vegetables.

**Wilt**
is caused by a fungus or bacteria that enters the water-carrying system of a plant. The vessels eventually block. Wilting is followed by browning and death of leaves and shoots. The whole plant will then die.

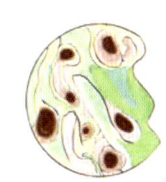

**Blight**
is a fungal disease that causes death of areas of plant tissue. There are different types of blight. Early blight, also known as target spot, is a common disease of tomatoes and potatoes that is usually first seen on foliage.

**Damping off**
is caused by numerous soil-inhabiting fungi. It affects the germinating seedling as well as the seedling itself, causing the stem to rot near the surface of the soil.

## SLUGS AND SNAILS

**Slugs and snails**
damage plants by licking and rasping leaves, young shoots and seedlings. They are often active at night and after rain.

## BUTTERFLY AND MOTH LARVAE

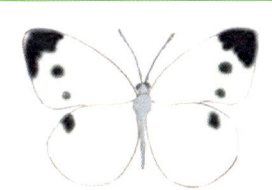

**Codling moth**
is a serious pest of apples, pears and English walnuts. The moth lays each of her eggs on a different fruit. The hatched larva chews through the skin or shell and feeds on the pulp. The caterpillar, when fully grown, will drop to the ground, usually with the fruit, and look for a place to pupate.

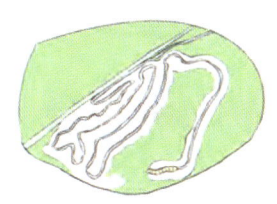

**Cabbage white butterfly**
lays its eggs on the underside of leaves that will feed the larvae and caterpillars. They prefer all members of the cabbage family, including kale, broccoli and cauliflower.

**Leafminers**
are the larvae of tiny moths. The larvae feed beneath the surface of a leaf, leaving a distinctive silvery trail. The adult caterpillar falls from the leaf and forms its pupa in leaf litter and garden debris.

---

**petal** *see* flower

**petiole** *see* **leaf**

**pH**

The pH scale, from 0 to 14, measures how acidic or alkaline a substance is.

Litmus paper and various instruments are available for measuring the pH of garden soils. The pH of litmus paper ranges from 0 to 14, with 0 at the red end representing the most acidic, 14 (a dark blue) at the opposite end representing the most alkaline, and neutral in the middle.

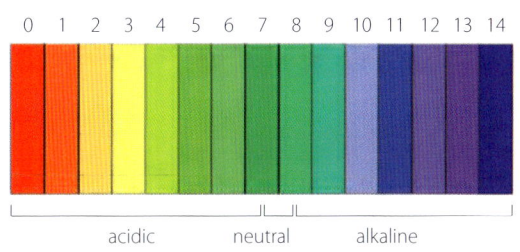

The colour scale key for measuring pH using litmus paper

*see also* **soil pH**

## phloem

Phloem is a cylinder of cells that makes up the inner bark of a tree.

It runs the length of the plant and transports nutrients from photosynthesis, mostly made in the leaves, to other parts of the plant.

PHLOEM

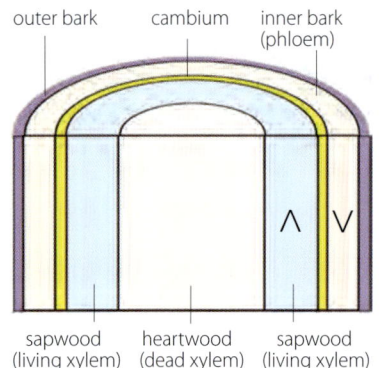

outer bark    cambium    inner bark (phloem)

sapwood (living xylem)    heartwood (dead xylem)    sapwood (living xylem)

Phloem (inner bark) carries sugars produced by photosynthesis to other parts of the plant

Xylem (sapwood) carries water and mineral salts from the roots to other parts of the plant

---

## phosphorus *see* plant nutrients

## photosynthesis

Photosynthesis is a process in green plants that uses energy from the sun to make carbohydrates from carbon dioxide in the air and water transported from the roots. It occurs mainly in the leaves of green plants.

A byproduct of photosynthesis is oxygen, that is released by the plant back into the atmosphere.

## physic garden

Physic gardens were established in Western Europe during the reign of the Emperor Charlemagne between 800 and 814 AD. He ordered all monasteries in his empire to plant a garden of healing herbs. These gardens were the pharmacies of the day.

The gardens were experimental. Monks in Europe and England learnt from ancient Greek, Roman and Arabic herbals and also studied plants that were not understood but were thought to have medicinal potential. Remedies of blended herbs or single herbs (simples) were used to treat patients being cared for in a monastery.

Physic gardens were the precursors of botanic gardens.

## piggyback plant

The piggyback plant is the ground cover *Tolmiea menziesii*. Mature leaves form plantlets at the base, where the petiole meets the leaf.

Plantlets can be separated and propagated.

piggyback plant
(*Tolmiea menziesii*)

## pinching *see* tip pruning

## pit composting *see* dig and drop composting

# plant kingdom

The plant kingdom is divided into groups of plants, from the most complex to the least complex.

*see also* **family**

PLANT KINGDOM

| | | | | | |
|---|---|---|---|---|---|
| **KINGDOM**<br>Plantae<br>(most complex) | angiosperms | algae | bryophytes | ferns | gymnosperms |
| **DIVISION**<br>vascular plants | angiosperms | bryophytes | ferns | fern allies | gymnosperms |
| **CLASS**<br>eudicots | Rosaceae | Geraniaceae | Lamiaceae | Vitaceae | Fabaceae |
| **FAMILY**<br>Rosaceae | *Prunus* | *Rosa* | *Malus* | *Rubus* | *Fragaria* |
| **GENUS**<br>*Prunus* | cherry | almond | apricot | blackthorn | |
| **SPECIES**<br>*Prunus avium* | *Prunus avium* cherry | | | | |

## plant nutrients

Plants need essential nutrients from the soil and the atmosphere to grow and flourish.

Carbon (C), oxygen (O) and hydrogen (H) are taken up from the atmosphere and from water. They account for about 94% of a plant's weight. The other 6% of essential nutrients come from minerals and organic matter. Of these, nitrogen (N), phosphorus (P) and potassium (K) are needed in larger quantities and are called macronutrients. Calcium, magnesium and sulfur are required in lesser amounts and are called secondary macronutrients. Micronutrients are required in very small amounts. These are: boron, chloride, cobalt, copper, iron, manganese, molybdenum, nickel, zinc and silicon.

Plants can only absorb soil nutrients in a chemical form dissolved in soil water.

A plant will exhibit deficiency symptoms if an essential nutrient is too low and toxicity symptoms if the level is too high. Generally, initial symptoms are expressed in new or older leaves. A balance of nutrients can usually be maintained by applying organic matter. Fertilisers are also often applied to correct these deficiencies.

| PLANT SOIL NUTRIENTS | | | |
|---|---|---|---|
| Nutrient | Why nutrient is needed | Deficiency symptoms | Excess symptoms |
| PRIMARY MACRONUTRIENTS | | | |
| Nitrogen (N) | Promotes green leafy growth, increases protein content, enhances nutrient uptake, essential for photosynthesis as it is a component of chlorophyll | Poor plant growth, plant light green, older leaves yellow, dying and falling early | Rapid weak growth, reduced flowering and fruit set |
| Phosphorus (P) | Essential for flower, fruit and root development, promotes photosynthesis and nutrient transport | Leaves dark green with red or purple blotches, poor fruit and seed set | Excessive amounts cause iron and zinc deficiency symptoms |
| Potassium (K) | Regulates photosynthesis and controls water uptake, promotes flowering, fruiting and lengthening of stems | Older leaves wilting, with scorched margins, yellow between the veins | Rare, affects the uptake of other nutrients but does not appear to be toxic to plants |
| SECONDARY MACRONUTRIENTS | | | |
| Calcium (Ca) | Stimulates root and leaf development | New leaves distorted or hook-shaped, blossom-end rot on some fruits like tomatoes and olives | Inhibits magnesium and boron uptake |
| Magnesium (Mg) | Essential for photosynthesis and forming proteins | Older leaves turn yellow, margins brown and curl | Rare, interferes with calcium uptake, spots on older, then on younger leaves |
| Sulfur (S) | Essential for protein synthesis and photosynthesis, promotes nodule formation in legumes | New growth yellow, eventually spreading to older leaves. In legumes decreases nodule growth resulting in less nitrogen. | Once a result of air pollution, causes symptoms of nitrogen deficiency |

| Nutrient | Why nutrient is needed | Deficiency symptoms | Excess symptoms |
|---|---|---|---|
| MICRONUTRIENTS | | | |
| Boron (B) | Essential for development of seeds, fruit and tissue | Reduced flowering, poor pollination and seed set, greatly reduced and distorted new growth, death of roots | Inhibits seed germination, yellowing of leaves, some tissue dying, premature leaf drop |
| Chloride (Cl-) | Regulates the release of water from leaves and balance of water and salts in cells, not to be confused with the gas chlorine | Wilting of leaf tips, distinct boundaries between yellowing and normal green colour on leaves | Smaller leaves, death of leaf margin tissue first on older leaves |
| Cobalt (Co) | Nitrogen fixing in legumes, plant growth and stem development | Inhibits development of nitrogen-fixing nodules in legumes | Young leaves yellow, inhibits seed germination and seedling growth |
| Copper (Cu) | Essential for photosynthesis and plant strength, required for nitrogen and carbon metabolism | Weak stems and branches, delayed flowering and sterility, young leaves wither and dry out | Inhibits iron uptake and seed germination, young leaves turn dark green and twist |
| Iron (Fe) | Promotes formation of chlorophyll, has a role in nitrogen fixation | Leaves pale green to yellow to white between the veins | Leaves turn bronze with possible specks of brown |
| Manganese (Mn) | Aids chlorophyll synthesis, accelerates seed germination | Young leaves pale yellow between the veins, with dark dead spots, stunted growth, poor bloom size | Leaves distorted with brown spots |
| Molybdenum (Mo) | Required in very small amounts, makes nitrogen available for protein synthesis, helps form legume nodules | Nitrogen deficiency, yellow blotches on older leaves between veins progressing to young leaves | Rare, symptoms of iron and cobalt deficiency |
| Nickel (Ni) | Important for seed germination | Seeds do not germinate | Inhibits seed germination and root/shoot growth |
| Silicon (Si) | Appears to improve plant response to stress and nutrient uptake | General deficiency symptoms remain undefined | Toxicity is rare in plants |
| Zinc (Zn) | Essential for plant growth hormones, good root development and chlorophyll synthesis | Leaves yellow between the veins, stems with short internodes and rosettes of small leaves | Darker than normal leaves |

## pleaching

Pleaching is a method of interweaving growing tree branches to form a high hedge or shaded arbour over a path. Shrubs can also be pleached.

## plug

Plugs are seedlings that have been germinated and grown in trays of small cells filled with a growing medium.

When the roots have grown enough, the plugs are pushed out of the trays and transplanted. The strong network of roots holds the soil of the seedling together. Coir trays with plugs can be cut up and the seedling planted together with the small coir cell.

A tray of coir cells with plugs

plug

A seedling removed from a cell with the soil held together by a network of roots

A plug also refers to a small division of a ground cover or grass, together with the soil it is growing in, that is used for propagation.

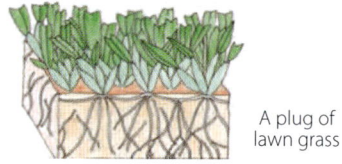

A plug of lawn grass

**plumule** *see* germination, seed

## pole cuttings

Pole cuttings are living stakes of hardwood taken from deciduous trees and shrubs that root easily.

Stakes are usually about 1–3 m long. In the dormant season, they are inserted in the ground where they will grow.

Plants like the Indian coral tree (*Erythrina variegata*), species of willow and bamboo can be propagated this way.

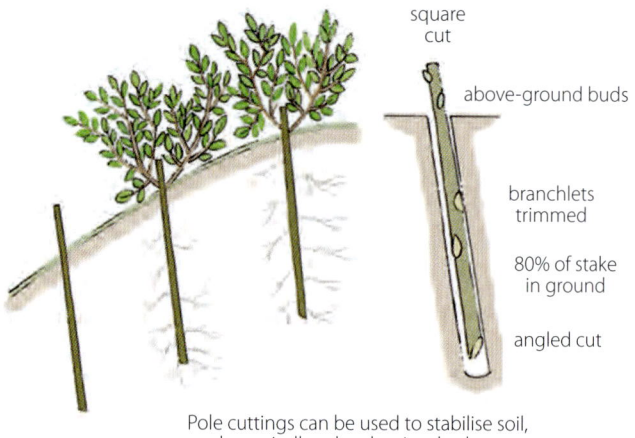

square cut

above-ground buds

branchlets trimmed

80% of stake in ground

angled cut

Pole cuttings can be used to stabilise soil, make a windbreak or begin a hedgerow

## pollarding

Pollarding involves cutting the main branches of a tree back to the trunk. This promotes growth of slender branches from the cut ends and a dense canopy of greenery.

Pruning of the new growth is done regularly.

Only some tree species can be pollarded, including elm and mulberry.

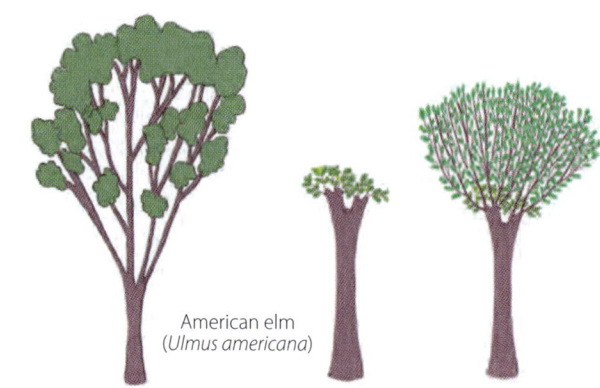

American elm (*Ulmus americana*)

## pollen

Pollen is minute grains, commonly yellow in colour, produced in the pollen sac of a seed plant. It is shed from sacs in the anther of a flower or from sacs on the male cone scales of gymnosperms.

## pollination

Pollination occurs in flowering plants (angiosperms) and non-flowering seed plants (gymnosperms).

In flowering plants (angiosperms), pollination is the transfer of pollen from the anther of a flower to the stigma of the same flower, or to the stigma of another flower. There are two types of pollination: self-pollination and cross-pollination. It is also possible to artificially pollinate flowers by hand. In non-flowering seed plants (gymnosperms), pollination is the transfer of pollen from a male cone to an ovule on the cone scale of a female cone of the same species. Pollen is commonly wind-borne.

Upon transfer, the pollen germinates to form a tube that transports the male sperm cell to the female egg cell, where the two unite (fertilisation). This occurs in both gymnosperms and angiosperms. After pollination and fertilisation seeds are formed.

Fruit trees like most nectarines, peaches, apricots and citrus that self-pollinate and bear fruit are said to be self-fruitful. Those that cannot produce fruit from their own pollen and need to cross-pollinate, like most sweet cherries, are said to be self-unfruitful. Pollination is an important factor when selecting and planting fruit trees.

*cf.* **fertilisation**

POLLINATION

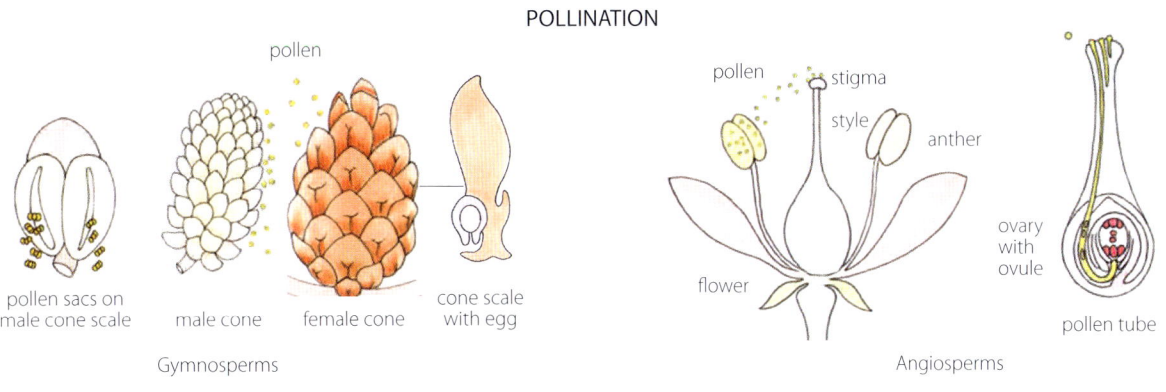

pollen

pollen sacs on male cone scale

male cone

female cone

cone scale with egg

Gymnosperms

pollen — stigma

style

anther

ovary with ovule

flower

pollen tube

Angiosperms

## pollinator

In angiosperms, a pollinator moves pollen from an anther to the stigma of a flower. In gymnosperms a pollinator, usually wind, moves pollen from the male cone to a female cone.

Insect pollinators include bees, wasps, ants, flies, moths, butterflies and beetles. Water can be a pollinator. Some vertebrates, like flying foxes are also pollinators.

Many orchids can only be pollinated by a specific insect pollinator.

Spider orchids attract specific wasp pollinators. The brown-clubbed spider orchid (*Caladenia phaeoclavia*) is pollinated by the wasp *Lophocheilus anilitatus*.

## polyculture

Polyculture is the practice of growing more than one plant species at a time, usually several, over a wide area or in a garden bed.

## polytunnel *see* greenhouse

## pome fruit

A pome is a fruit, like a pear or apple, that has a skin enclosing a fleshy layer, and a core with seeds.

Pomes are members of the apple subtribe Malinae in the rose family Rosaceae. They include edible fruits like apples, pears, quinces, loquats and crab apples, and ornamental pears and quinces. There are hundreds of pome tree cultivars.

Trees are deciduous and dormant in winter, flower in early spring, and bear fruit in summer.

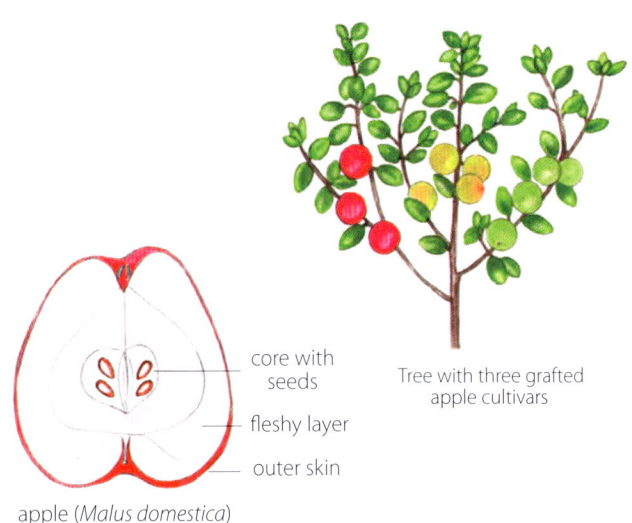

core with seeds

fleshy layer

outer skin

apple (*Malus domestica*)

Tree with three grafted apple cultivars

**pot bound** *see* root-bound

**potager**

Potager is a French term for an ornamental vegetable or kitchen garden.

**potash**

A former name for potassium.

**potassium** *see* plant nutrients

**potting mix** *see* medium

**potting up**

To transplant a seedling into a pot, or a potted plant into a larger pot or container.

**powdery mildew** *see* pests and diseases of gardens

**pricking out**

A seedling is pricked out by using a dibbler to loosen the soil and gently lift it without damaging the roots. This is done when the first set of true leaves are established.

When transferring the seedling to a pot, it is held by a leaf to minimise damage to the stem. A hole is made in the soil using a dibber, and the seedling is placed in it almost to the base of the leaves. The soil is firmed down and watered in with a fine spray.

true leaves

seed leaves

dibber stick

**primocane** *see* bramble

**propagation**

The reproduction of plants by any number of natural means, like seeds, offsets and bulbs, or by artificial means, like grafting, cuttings and layering. The two methods are vegetative propagation and sexual propagation.

**propagator**

A propagator is used for growing seedlings and cuttings in trays. The clear cover maintains humidity, and has adjustable air vents.

Some propagators have a misting chamber in the base. The tray of plants is suspended over the mist so that the roots are kept moist and growth is quicker.

A propagator can be used indoors but requires overhead lighting once the seedlings have emerged to avoid legginess.

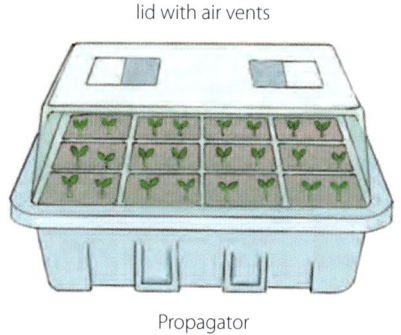

lid with air vents

Propagator

## propagule

A propagule is a structure capable of producing a new plant.

In flowering plants, propagules include seeds, bulbs and runners. In non-flowering plants, like ferns and bryophytes, examples include spores and rhizomes.

## protists

Protists are mostly unicellular organisms that are part of the soil food web.

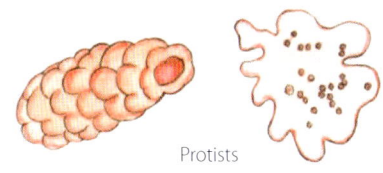

Protists

They are predators and are the main consumers of bacteria and fungi. Protists 'recycle' bacteria and fungi by eating them. At the same time, they control the size of these communities.

## pruning *see* next page

## pseudobulb

A pseudobulb is the enlarged portion of an orchid flower stem.

They are generally found in epiphytic orchids, and serve as water and nutrient storage organs. They grow from the orchid's rhizome. All leaves and inflorescences arise from a pseudobulb.

Pseudobulbs eventually die away and become back bulbs.

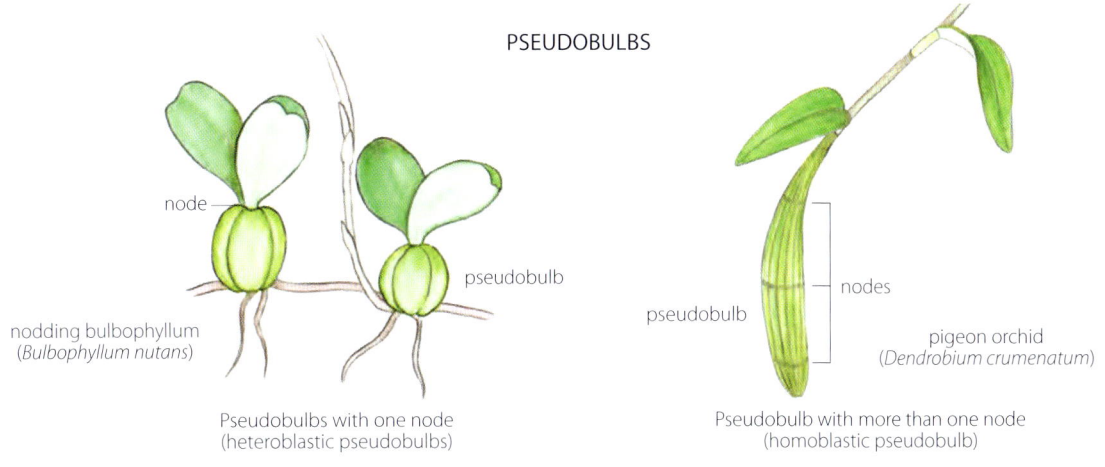

PSEUDOBULBS

node

pseudobulb

nodding bulbophyllum
(*Bulbophyllum nutans*)

Pseudobulbs with one node
(heteroblastic pseudobulbs)

nodes

pseudobulb

pigeon orchid
(*Dendrobium crumenatum*)

Pseudobulb with more than one node
(homoblastic pseudobulb)

## pup *see* offset

## pyrethrum

The yellow centre of the pyrethrum daisy (*Tanacetum cinerariifolium*) contains the chemical pyrethrin which is toxic to insects. It has been used for centuries as an insecticide. Today, the pyrethrin is extracted and put in sprays to kill insects like moths, leafhoppers, ants and aphids. It is harmless to people and breaks down quickly.

If sprayed directly onto the offending pest, it will die almost instantly. However, pyrethrums are also highly toxic to good insects like honey bees, and to fish, lobster, oysters and aquatic insects.

Synthetic pyrethrums, called pyrethroids, are stronger but also more toxic than natural pyrethrins.

## pruning

A plant is pruned to train it or to maintain its shape. Any dead or diseased material is also removed. Pruning encourages new growth, more flowers, and in fruit trees, more or better quality fruit.

There are five main pruning techniques: tip pruning, heading back, thinning, renewal pruning and rejuvenation pruning. Shearing, another method, cuts most of the terminal shoots when hedging or maintaining a topiary.

Pruning techniques differ according to the plant. Bush tomatoes and climbing tomatoes require different techniques. Depending on the species, there are several ways to prune a grape vine.

Fruit trees flower and fruit each year, but some do so only on new branches or specific parts of older branches. Peaches (*Prunus persica*) fruit mainly on 1-year-old wood, and cherries (*Prunus avium*) bear fruit on 1-year and older wood, and on long-lived spurs. Citrus trees bear fruit on the current season's new growth, and olives trees bear fruit only on 1-year-old wood.

A properly pruned tree or shrub will seal wounds naturally. It is no longer considered helpful to apply a sealant.

PRUNING

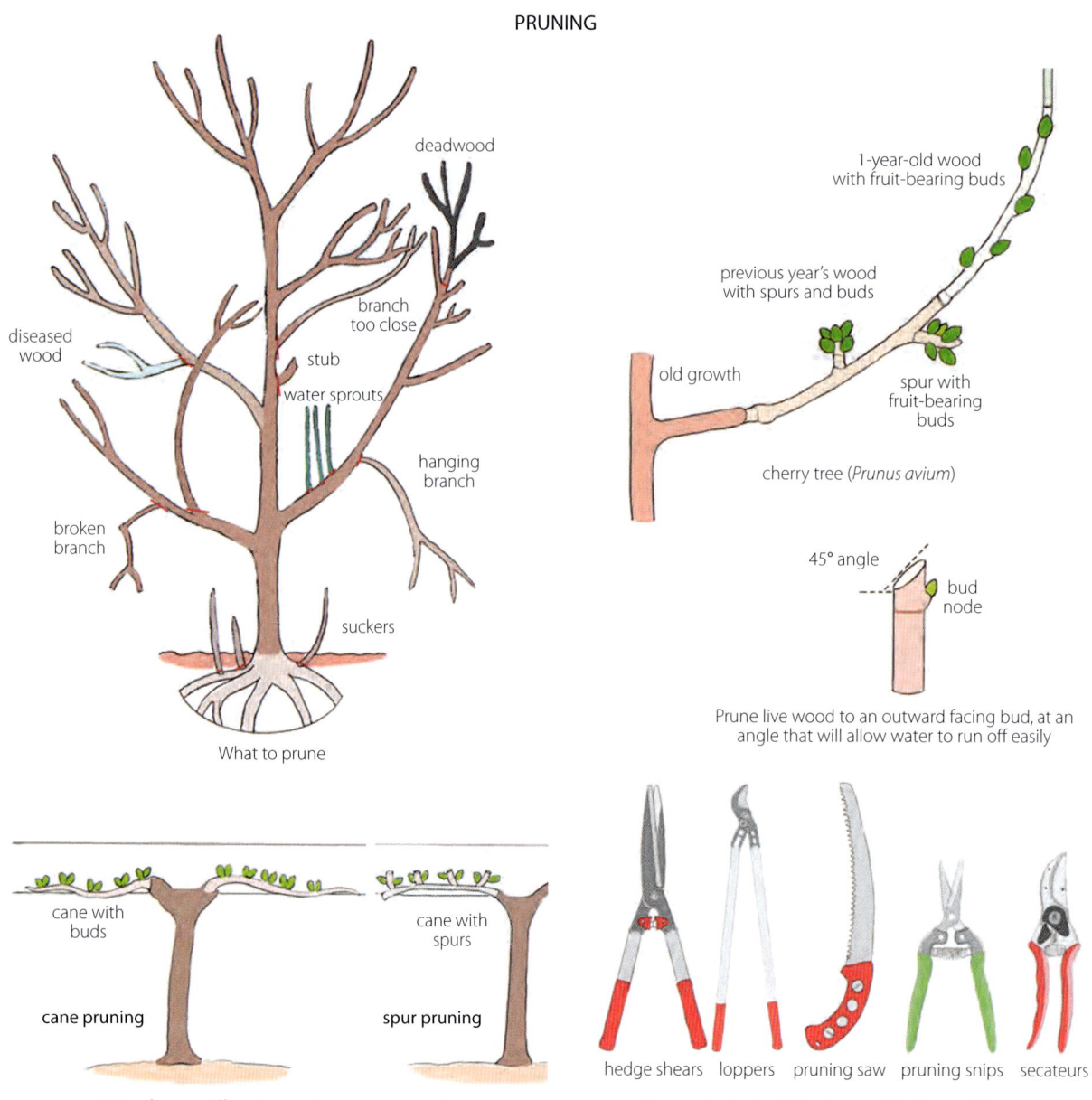

deadwood

branch
too close

diseased
wood

stub

water sprouts

hanging
branch

broken
branch

suckers

What to prune

1-year-old wood
with fruit-bearing buds

previous year's wood
with spurs and buds

old growth

spur with
fruit-bearing
buds

cherry tree (*Prunus avium*)

45° angle

bud
node

Prune live wood to an outward facing bud, at an
angle that will allow water to run off easily

cane with
buds

cane with
spurs

**cane pruning**

**spur pruning**

Two of many different ways to prune a grape vine

hedge shears    loppers    pruning saw    pruning snips    secateurs

Pruning tools should be thoroughly cleaned after use

## qanat

Throughout many arid regions of the world, gardens, agriculture and permanent settlements are supported by the ancient qanat system. Qanats tap alluvial aquifers at the head of valleys and conduct the water along underground tunnels by gravity, often over many kilometres, to a holding basin. Water is distributed  for irrigation through a network of dams, gates and channels.

It is believed to have originated in Persia in the first millennium and today is in use in more than 34 different parts of the world, including Iran, Afghanistan, North Africa, Central Asia and China.

QANAT

rainfall

valley

mother well

surface runoff protection mound

access shaft

desert

water table

irrigated oasis

bedrock

aquifer

alluvial fan

water collection

water transport

Water is distributed through a network of dams, gates and channels

gently sloping main channel

---

**radicle** *see* germination, seed

## raised garden beds

Raised garden beds are rectangular boxes filled with soil and made of untreated wood, PVC, or more permanent materials like galvanised iron and brick. They are mostly about 2 m long, but can be any length; 30 cm high, but can be waist height; and no wider than 1.2 m so that it is easy to reach inside the bed.

A popular way to make soil for a raised garden bed  is to first spread overlapping cardboard and newspaper on the base. Organic 'brown' layers of leaves, straw and coir are then alternated with nitrogen-rich 'green' layers of seedless grass clippings, coffee grounds and manure. These two layers can be repeated to achieve the required depth. Finally, there is a compost layer that is topped with mulch. Each layer should be thoroughly watered. The bed will settle as the layers break down.

RAISED GARDEN BEDS

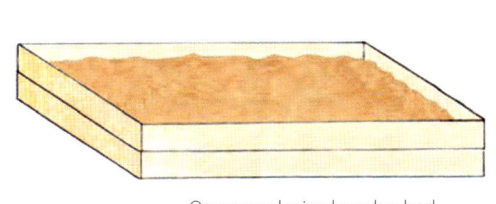

On-ground raised garden bed

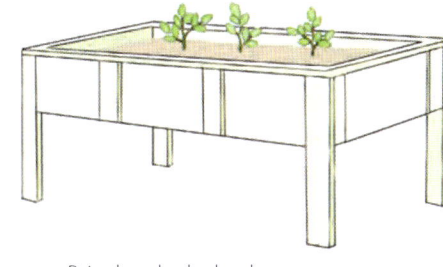

Raised garden bed on legs

Root depth of vegetables can help determine the best height for raised garden beds

| Shallow rooting (30–45 cm) | Medium rooting (45–60 cm) | Deep rooting (60–90 cm) |
|---|---|---|
| basil | beets | asparagus |
| bok choy | bush beans | beans (lima) |
| coriander | cantaloupe | cardoon |
| lettuce | carrots | okra |
| mint | cucumbers | parsnips |
| onions | daikon | pumpkins |
| oregano | egg plant | rhubarb |
| radishes | peas | squash (winter) |
| shallots | peppers | tomatoes |
| spinach | rosemary | watermelon |
| strawberries | sage | |
| tarragon | squash (summer) | |
| thyme | turnips | |

An example of layers for making soil

| Mulch | 5 |
|---|---|
| Compost layer | 4 |
| **Brown layer** of organic matter like leaves, coir and straw, and wet thoroughly | 3 |
| **Green layer** of nitrogen-rich matter like grass clippings and manure, and wet thoroughly | 2 |
| **Cardboard and newspaper**, well watered, to suppress weeds | 1 |

## rejuvenation pruning

To completely renew growth on a shrub, all stems are cut down to ground level.

This is the easiest way to prune large, overgrown lilacs and flowering quinces that have been left untouched for a number of years. New shoots will develop in the next growing season.

All stems are cut down to near ground level

## rambling roses

Rambling roses are vigorous shrubs with long, flexible stems that grow from the base of the plant. They are bigger than climbing roses and need more space.

They are easy to train over trellises and pergolas, or they will scramble through bushes and trees or over objects.

*see also* **roses**

## relative humidity

Relative humidity is the percentage of water vapour in the air relative to the highest amount, 100 percent, it can hold. After this, the water vapour will condense and form dew.

Warm air has a higher moisture holding capacity than cooler air. Air at 21°C (70°F) will hold twice as much moisture as air at 10°C (50°F).

## renewal pruning

To renew growth on a shrub, about one third of the stems are cut down to ground level each year.

The oldest stems are chosen for removal. This makes the shrub more open and encourages new growth from the base.

Deciduous shrubs like rhododendrons and azaleas respond well to renewal pruning.

One third of older stems are cut down to ground level

## repair grafting

Repair grafts restore trees that have been damaged.

The most common grafts used are bridge grafting and inarching.

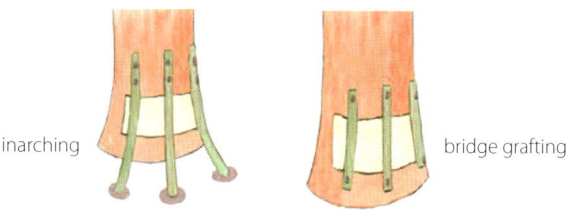

inarching

bridge grafting

## rhizome

A rhizome is a stem that usually grows horizontally underground, but it can also grow vertically. Some grow above the ground, like those of some ferns. A rhizome enables a plant to spread.

Rhizomes send out roots and shoots from nodes. Pieces of rhizome with a node and roots can be used to grow clones of the parent plant. Unlike stolons, rhizomes are adapted for food storage and as organs of survival during dormancy.

Climbing ferns and some creeping ferns have above-ground rhizomes. Other plants, like some species of viola and primrose, have a vertical rhizome.

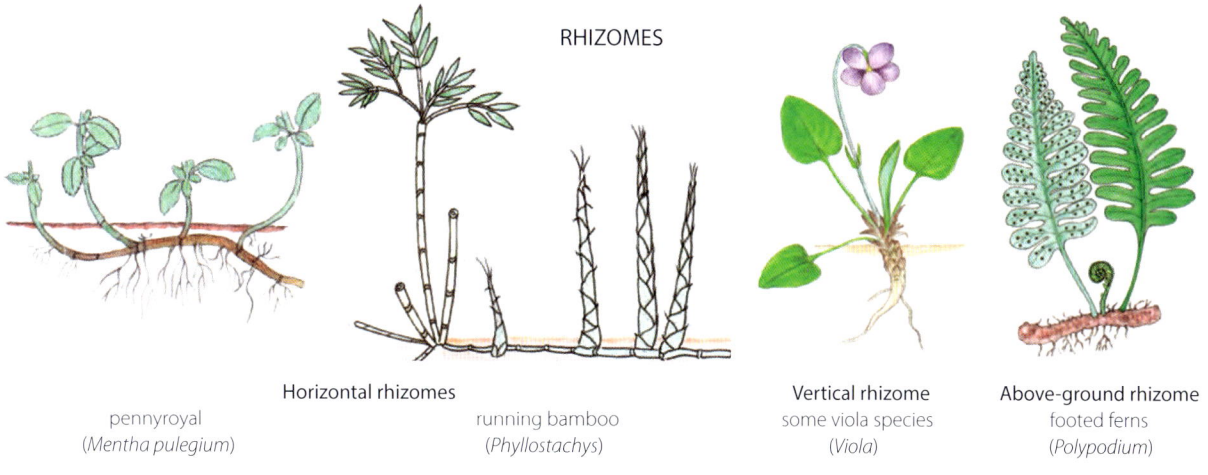

RHIZOMES

Horizontal rhizomes

pennyroyal
(*Mentha pulegium*)

running bamboo
(*Phyllostachys*)

Vertical rhizome
some viola species
(*Viola*)

Above-ground rhizome
footed ferns
(*Polypodium*)

## ring barking

Ring barking is the removal of a strip of bark from around the trunk or branch of a tree. Ring barking is a way of killing a tree.

During the process the cambium and phloem layers are removed, blocking the flow of nutrients from the foliage downward. The xylem is also removed, blocking nutrients that flow upward through the roots.

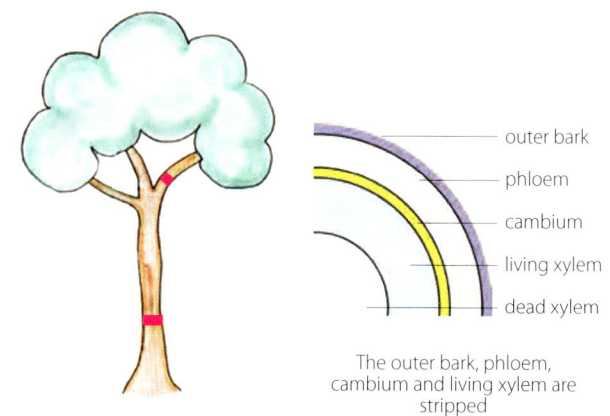

outer bark
phloem
cambium
living xylem
dead xylem

The outer bark, phloem, cambium and living xylem are stripped

## rockwool

Rockwool is a man-made mineral fibre.

It is formed from basalt rock and chalk that have been melted and spun until they become a fibre, much like fibreglass. These fibres are then compacted and shaped into a slab or smaller cubes for raising seedlings. It is a popular soilless medium in hydroponics.

## root cuttings

A range of herbaceous perennials, like acanthus, oriental poppy and Japanese anemone, can be propagated from root cuttings.

Root cuttings are also used to propagate plants that naturally produce suckers. A sucker with roots is separated from the parent and either planted directly into garden soil or grown on in a pot. Plants that give rise to suckers include lilacs, red raspberries, blackberries and Japanese quince.

A sucker grows from a bud on the parent root

A new plant from the separated sucker

## root pruning

Root pruning is the trimming of a plant's roots.

Tree roots are pruned prior to transplanting to prevent stress. Young trees are usually pruned several months before transplanting by cutting through the roots at the drip line of the plant. This allows new roots to form nearer the trunk.

Indoor plant roots are trimmed to check their growth, and bonsai roots are pruned to stunt growth.

## root tuber *see* tuberous roots

## root-bound

A root-bound plant has outgrown its container. The roots have nowhere to grow and become entangled or coiled around the inside of the pot.

The roots of these plants are pruned and loosened before repotting or planting out in the garden.

## rooting hormone

A rooting hormone is a chemical that stimulates root growth.

Using a root-stimulating hormone increases the success rate of some cuttings. The hormone comes in powder, liquid or gel form, and is applied to the cutting before planting in a growing medium.

## roots

A root is the usually underground part of a plant that anchors it and absorbs water and nutrients from the soil.

A taproot is a single enlarged root, like a carrot or a turnip. Some taproots have well-developed lateral roots.

Fibrous roots are a mass of roots of similar size. They are typical of grasses, palms, corms, and bulbs like daffodils and onions. Monocotyledons have fibrous roots, and eudicots have taproots.

Aerial roots grow above ground, as do those of strangler figs and epiphytic orchids.

About 90% of all land plants have an association, called a mycorrhiza, between their roots and fungi in the soil. Roots provide the fungi with nutrients from photosynthesis, and fungi provide plants with nutrients in the soil that they would otherwise be unable to access. They also store water for the plant to use in drier times.

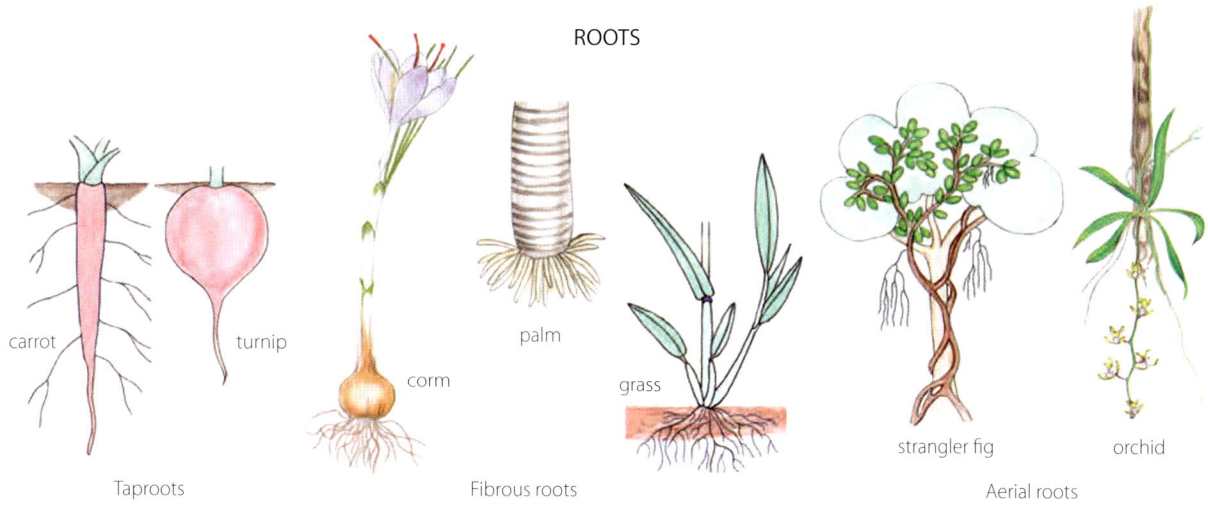

ROOTS

carrot  turnip

corm

palm

grass

strangler fig  orchid

Taproots  Fibrous roots  Aerial roots

## rootstock

A rootstock is the plant used for the root system of a graft.

The scion from the plant to be propagated is inserted into the rootstock so that the two grow together and unite to form a single plant.

A scion can be grafted onto a rootstock that suits local growing conditions, allowing. cultivars, like the Granny Smith apple, to be grown in different climates.

Specific rootstock can also be chosen to produce a standard or dwarf tree, provide resistance to disease or to suit a particular type of soil.

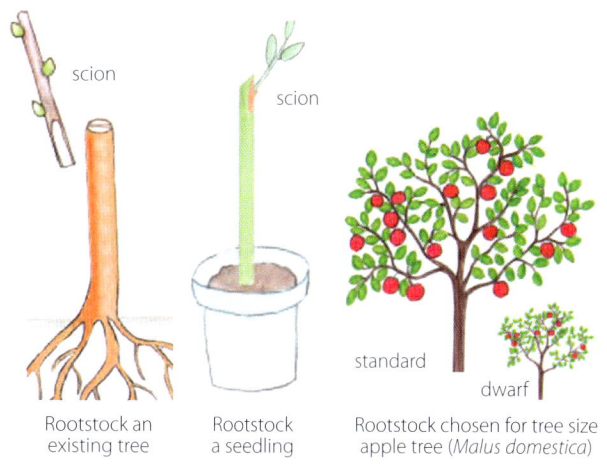

scion

scion

standard

dwarf

Rootstock an existing tree

Rootstock a seedling

Rootstock chosen for tree size apple tree (*Malus domestica*)

## roses

Roses are perennial, erect, trailing or climbing shrubs, usually with prickly stems. They are usually deciduous and are sold bare-rooted in winter. At other times of the year they are sold in pots and can be planted at any time.

There are about 250 species of wild roses and thousands of hybrids resulting from crosses between them.

Most horticulturalists recognise three groups of roses: wild species roses, old garden roses that are hybrids that existed before 1876 and modern hybrid roses that were not in existence before 1867.

Species roses have been grown for centuries. They are hardy, usually bloom once a year, and have simple flowers with 4 to 12 petals. Their colour is mainly pink or white. They are crossed to make old garden rose hybrids and modern rose hybrids.

Old garden roses may have an initial spring bloom and produce no more flowers for the rest of the year. Their beauty often lies in their heavily fragrant flowers and clusters of colourful hips in autumn. The species damask roses are regarded as the ancestors of most European hybrids. The two basic forms of the damask rose derive from crosses with *Rosa gallica*: *Rosa gallica* x *R. Phoenicea* and *R. gallica* x *R. moschata*.

In 1867, the first hybrid tea rose, La France, was introduced by the French breeder Guillot. It was a cross between a repeat blooming hybrid perpetual and a free-flowering tea rose. La France had large, repeat-blooming flowers with 30 to 50 petals. The most famous hybrid tea is the 'Peace' rose introduced in 1945. The hybrid tea was an instant success and began the era of modern rose breeding. Hybrid teas are the most popular modern rose and many thousands have been bred and introduced to the market.

Modern hybrid roses are showy repeat bloomers. They include shrub roses, climbers and ramblers, floribundas, miniature roses, weeping roses, ground cover roses and David Austin roses (also known as English roses).

*see also* **climbing roses, rambling roses, sport**

## ROSES

Rose leaves are divided into an uneven number of leaflets

Many species roses and old garden roses are admired for their beautiful leaves and colourful hips

stamens and stigmas

bud

hypanthium

'Seeds' (achenes) develop in the hypanthium (hip)

Species rose
red-leaved rose (*Rosa glauca*)

Species roses are typically pink with 5 petals, and bloom only once in a season

| Species roses include | Old garden roses | Modern roses |
|---|---|---|
| *Rosa banksiae* (Banks' rose) | Gallica | Hybrid Tea |
| *Rosa chinensis* (China rose) | Alba | Pernetiana |
| *Rosa glauca* (Red-leaved rose) | Damask | Polyatha |
| *Rosa canina* (Dog rose) | Centifolia or Provence | Floribunda |
| *Rosa gallica* (French rose) | Moss | Grandiflora |
| *Rosa moschata* (Musk rose) | Portland | Miniature |
| *Rosa phoenicia* (Phoenician rose) | China | Climbing and Rambling |
| *Rosa blanda* (Smooth rose) | Tea | Shrub roses |
| | Bourbon | David Austin roses (also called English roses) |
| | Noisette | Canadian hardy |
| | Hybrid perpetual | Ground cover |
| | Hybrid musk | Patio roses |
| | Hybrid rugosa | |

**runner** *see* **stolon**

## saddle grafting

In saddle grafting, a wedge-shaped cut is made on top of the rootstock, and an inverted v-shaped notch is cut into the shoot (scion) to match the rootstock. The scion is pushed on to the rootstock to make a secure join and bound with tape.

The rootstock and scion should have the same diameter.

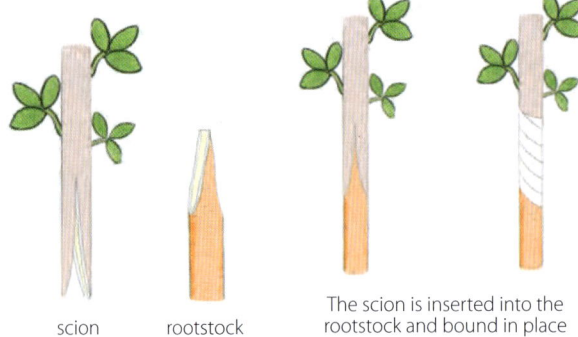

scion    rootstock

The scion is inserted into the rootstock and bound in place

## salinity

Salinity refers to the concentration of salts in water or soil, particularly when the levels become high and harmful to plant growth. This is a worldwide problem with many causes, including clearing long-lived vegetation in drier areas. Runoff from these soils causes salinity in rivers and streams.

## sand

Sand is small, loose single grains of disintegrated rock. It is well-aerated and free-draining but has no capacity to hold water or nutrients. Sand particles range in size from 0.05–2.0 mm, silt from 0.002–0.5 mm and clay < 0.002 mm.

## sandy soil

Sandy soil has a gritty texture. Its large particles separate easily, so there are no pockets for holding water and nutrients. Adding organic matter and compost to the sandy soil helps increase water retentiveness.

Cover crops like crimson clover (*Trifolium incarnatum*) and alfalfa (*Medicago sativa*) can be chopped and dropped before seeds form to provide organic matter and nitrogen. Herbs like rosemary, lavender and oregano, and root vegetables like carrots, grow well in sandy soils.

## sap

All plants are dependent on sap, the fluid that contains dissolved mineral salts, sugars and other nutrients that are circulated to various plant tissues.

It may be clear, like the sap of sugar cane and sugar maples, or it may be milky, like the sap of the rubber tree. The sap of many plants is poisonous. There are over 2000 species of *Euphorbia*, commonly called spurges, that have white, milky sap that is an irritant.

*see also* **pests and diseases of gardens**

## sapling

A sapling is the stage between a seedling and a mature tree.

It has a slender trunk with a diameter of 5–10 cm and a height up to 140 cm above ground level.

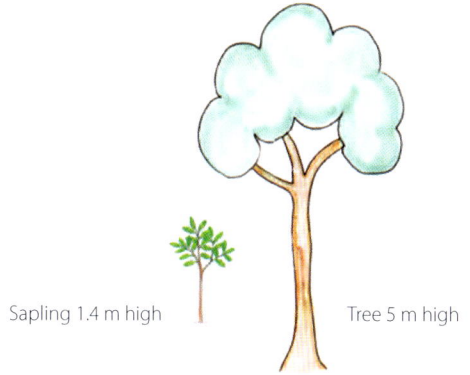

Sapling 1.4 m high          Tree 5 m high

**sapwood** *see* xylem

**scale** *see* honeydew, pests and diseases of gardens

**scalping** *see* thatch

**scaly bulb** = imbricate bulbs

## scarification

Scarification is a natural process that removes enough of a hard seed coat to make it permeable to water so that it will germinate.

In nature, scarification may occur during winter when the soil is frozen, or by fire, fungi and microbes in the soil, or when the seed passes through the gut of an animal. Scarification can be forced artificially by using an abrasive material, like sandpaper on the seed coat, or an instrument, like a sharp knife. It can also be done chemically or by using hot water.

### scientific name

The scientific name, also called the botanical name, is the formal name of a plant accepted by the scientific community, as opposed to its common name.

It has two parts. The first is the genus, or group, of plants to which it belongs. The second is its specific, or species, name. *Malus domestica* is the scientific name for what is commonly called an apple.

**Scientific name**
Genus:      *Malus*
Species:    *domestica*

**Common name**    apple

### scion

A scion is any part taken from one plant and inserted into a different plant with roots (the rootstock), so that the two grow together and unite to form a single plant.

The scion is a single bud on a piece of bark, with or without some wood attached, or a twig with two or more buds. It is taken from a variety, species or cultivar for propagation.

A scion with one bud is used for 'budding' and a scion with two or more buds is used for grafting.

SCION

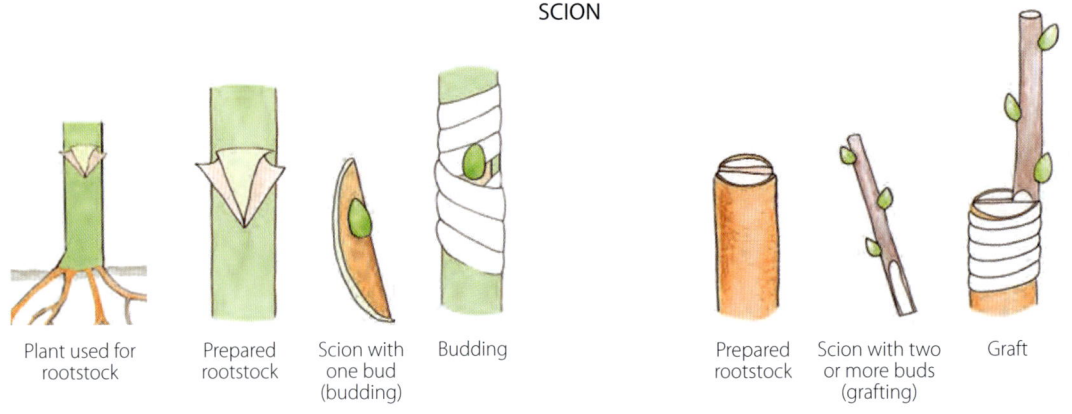

| Plant used for rootstock | Prepared rootstock | Scion with one bud (budding) | Budding | Prepared rootstock | Scion with two or more buds (grafting) | Graft |

### scooping

Scooping is a method of propagating bulbs used primarily with hyacinths (*Hyacinthus*).

It is done when the bulb is dormant. The base of the bulb is scooped out to expose the leaf scale bases, with the rim left intact. Fungicide is applied to the cut surface. The bulbs are placed in moist, coarse sand to half their depth, with the scooped base facing upwards, and kept moist in a warm dark place.

After about 12 weeks, bulblets form in the scooped-out area at the base of the leaf scales. Once new roots form on the developing bulblets, they can be potted up.

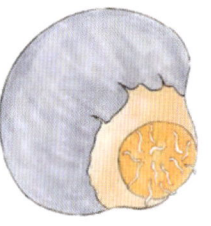

Basal plate with roots

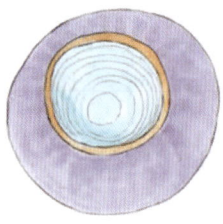

rim of basal plate left intact

Basal plate scooped

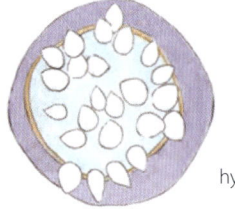

hyacinth (*Hyacinthus*)

Basal plate with bulblets

## scoring

Scoring is a method of propagating bulbs used primarily with hyacinths (*Hyacinthus*) and the daffodil family (*Narcissus*).

At the end of dormancy, the basal plate is scored across with a sharp, sterile knife. The bulbs are placed in a dry medium or on a wire tray in a warm, dark place, with the scored side facing upward. When the cuts open up, they are dusted with fungicide. The bulbs are kept moist by occasional spraying with water.

After about 12 weeks, bulblets should have formed on the cuts. Bulblets can be potted up in autumn ready for planting out in the spring.

Basal plate with roots

Basal plate scored

Basal plate with bulblets

hyacinth (*Hyacinthus*)

## scramblers

A scrambler is a plant with long stems and a sprawling, climbing or creeping habit. Their prolific growth enables them to scramble over large areas, even over shrubs and trees and nearby structures. Star jasmine (*Trachelospermum jasminoides*) with its fragrant flowers, scrambles vertically in this way, or horizontally as a ground cover.

## seaweed

Seaweed is classified into three broad groups, brown seaweed, red seaweed, and green seaweed. It can be wild-harvested or farmed. There are sometimes laws preventing the collection of natural seaweed from coastal environments.

Diluted extracts of seaweed are applied to soil or used as a foliar spray to promote growth, prevent pests and diseases, and improve nutrient uptake. It also enhances seed germination and improves flower set and fruit production. Untreated seaweed is beneficial in compost and as a top dressing.

Brown seaweed
kelp
(*Macrocystis*)

Red seaweed
laver
(*Porphyra*)

Green seaweed
spaghetti algae
(*Chaetomorpha*)

## seed

A seed is a result of the fusion of a male sex cell from pollen with a female sex cell in the ovary of a flower or on the female cone scale of a gymnosperm. Given the right conditions, a seed can grow into a new plant with characteristics from both parents.

Seed-bearing plants belong to two groups: flowering plants (angiosperms) and non-flowering plants (gymnosperms). In angiosperms the seed is enclosed in a fruit, and in gymnosperms the seeds are exposed, usually on the cone scale of a female cone.

All seeds have a seed coat, one or more cotyledons, and an embryonic tiny plant. The seed coat is protective. Cotyledons store food for when the seed germinates and begins to grow. The embryo has a plumule that will become the above-ground stem with leaves and buds, and a radicle that will grow into the root system.

## SEEDS

Flowers
produce seeds

plumule —
— embryo
radicle —
— seed coat
— cotyledon (seed leaf)

Seed

broad bean seed (*Vicia faba*)

Fruit encloses
seeds in a pod

cone
scale

Female cone
produces seeds

Winged seeds
are exposed
on a female
cone scale

pine (*Pinus*)

Conifer

**Seed bearing plants with flowers
(angiosperms)**

**Seed-bearing plants that lack flowers
(gymnosperms)**

**seed coat** *see* seed

**seed leaf** = cotyledon

**seed potato**

A seed potato is a small potato tuber that can be planted whole or a larger potato cut into smaller pieces, each with at least one vigorous bud (eye). The new plant will be a clone of the parent plant.

**seed raising mix** *see* medium

**seed saving**

The best seeds to save for the following season are those that grow true to type.

True to type seeds will reliably reproduce the parent plant. These include heirloom vegetable varieties and self-pollinating annuals like beans, peppers, peas and tomatoes. Most annual flowers are prolific seed-bearers and are easy to collect seed from for the next year. There are different techniques for saving seeds.

Plants with separate male and female flowers, like corn and cucumbers, can cross-pollinate and should be carefully hand-pollinated for reliable seeds. All brassicas can cross-pollinate with one another if they are in the same garden and will produce unreliable seeds.

### SAVING SOY BEAN SEEDS

Soybeans (*Glycine max*) are self-pollinating
and seeds reliably reproduce the parent plant

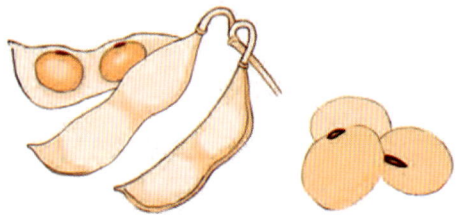

Harvest the seeds when they are hard
and the pods are dry and brittle

Depending on conditions, store seeds in an
airtight jar or a jar covered with cheesecloth

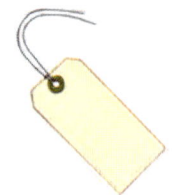

Tag seeds with name and date
of collection.
Seeds have a germination rate of
about 80% for the first year.

## seed tray with cells

Seed trays with cells are used for growing seedlings from seeds.

Once planted, they are usually placed in a propagator, greenhouse or hot house. Coir or paper seed trays can be cut and the cells planted together with the seedling.

Reusable plastic seed tray

Coir cells can be cut and planted as plugs directly into the garden

## seed tuber = seed potato

## seedling

A seedling is a germinated seed that has grown and has its first true leaves and roots.

Seedlings need moisture and warmth to grow. If they are crowded together, they are thinned out so that those that remain can grow without competition.

If grown inside, work areas, pots, trays and tools should be thoroughly cleaned to avoid disease. A greenhouse or propagator provides a humid, stable environment for seedlings to grow. Seedlings will need overhead light, or they will lean towards a light source, and the stems will elongate (etiolate) and weaken. Heat pads are used below trays to encourage root growth without changing the air temperature around the plant.

When they are large enough to handle, preferably by the leaves, they are pricked out and transferred to individual pots or seed trays with cells. Holes are made in the soil with a small dibber, and the roots are inserted and gently firmed in place. When there is sufficient growth, the seedlings are hardened off in a cold frame until they are ready to be planted out in garden beds.

## selective breeding

Humans selectively breed plants to develop new cultivars with desirable characteristics like drought tolerance, improved fruit flavour or resistance to disease.

Traditionally, selective breeding is done by propagating from a plant chosen for its exceptional qualities. Others are a result of cross-breeding two closely related plants with different useful traits.

Over hundreds of years, growers have been breeding wild mustard (*Brassica oleracea*) to produce many different cultivars.

### SELECTIVELY BRED CULTIVARS OF WILD MUSTARD

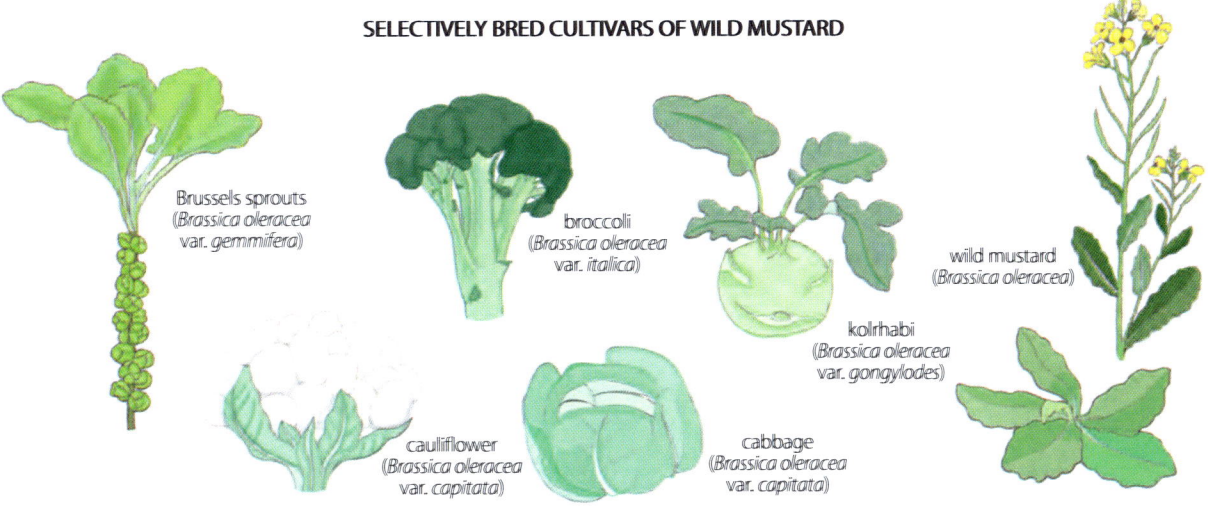

Brussels sprouts
(*Brassica oleracea*
var. *gemmifera*)

broccoli
(*Brassica oleracea*
var. *italica*)

wild mustard
(*Brassica oleracea*)

kolrhabi
(*Brassica oleracea*
var. *gongylodes*)

cauliflower
(*Brassica oleracea*
var. *capitata*)

cabbage
(*Brassica oleracea*
var. *capitata*)

## self-fertiisation

Self-fertilisation ocurs between flowers on the same plant or within a single flower. The sperm cell in the pollen of a flower unites with the egg cell in the same flower, or the egg cell in a flower on the same plant.

## self-fruitful *see* pollination

## self-pollination

In flowering plants, self-pollination occurs when the pollen from an anther is deposited on the stigma of the same flower, another flower on the same plant, or the flower of another plant of the same variety.

Some flowers that self-pollinate require pollen from a different flower. One way this is ensured is for the male pollen and female stigma to mature at different times. Other flowers, like some orchids, self-pollinate before the flower opens.

Self-pollination keeps the genetic characteristics of the plant stable, but cross-pollination allows for genetic diversity and the ability to adapt to change.

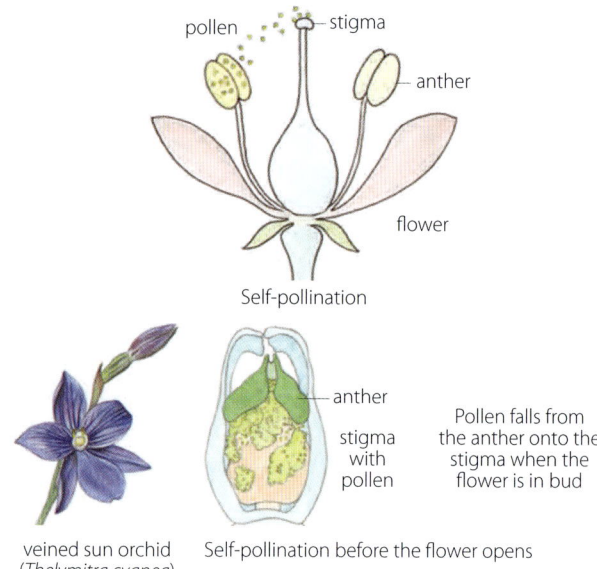

Self-pollination

veined sun orchid
(*Thelymitra cyanea*)

Self-pollination before the flower opens

Pollen falls from the anther onto the stigma when the flower is in bud

## self-unfruitful *see* pollination

## semi-hardwood cuttings

Semi-hardwood cuttings, also called semi-ripe cuttings, are taken from a stem between the softwood and the woody stages. It is spring growth that has started to mature.

Cuttings are made below a node or as a heel cutting, and the growing tip and lower leaves are removed. Larger leaves may be cut in half to minimise transpiration. The cut end is dipped in rooting hormone and inserted into a potting mix. It needs to be kept in a moist, light environment while taking root.

softwood

semi-hardwood

hardwood

rooting hormone

Semi-hardwood cuttings need a warm, moist atmosphere to take root

## semi-ripe cutting = semi-hardwood cuttings

## sepal *see* flower

## separation

Some naturally occurring parts of a plant can be separated from the parent and propagated.

These include bulblets, offsets, keikis, and tillers of grasses and sedges.

*cf.* division

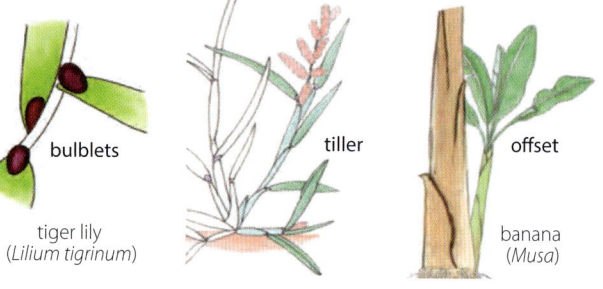

bulblets

tiger lily
(*Lilium tigrinum*)

tiller

offset

banana
(*Musa*)

## serpentine layering

Serpentine layering is a method of propagation. A flexible stem is bent over to form a series of loops. The troughs of the loops are buried in soil, and the crests are exposed. Each underground node is wounded to promote root formation.

Once roots have developed, the loops are severed to produce individual plants.

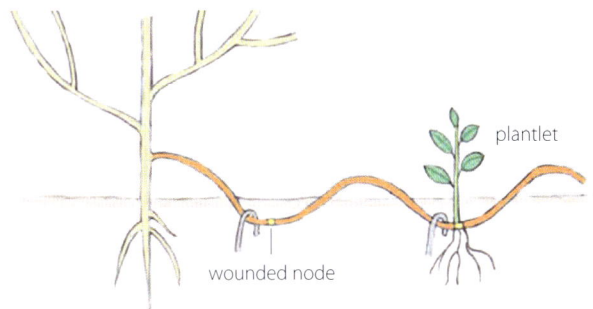

This technique suits climbers with long flexible stems like clematis, wisteria and grapes

## set

A set is a small, immature onion bulb grown from seed sown the previous year.

Sets are the most common way of propagating onions. They are available from nurseries.

## sett

Rhizomes can be broken up into pieces, called setts, for propagation. Each sett has one or more buds.

Plants propagated by setts include turmeric and ginger.

rhizome
iris (*Iris*)
setts

## sexual propagation

The reproduction of plants by seeds.

## shade house

A shade house is a garden structure that provides protection and suitable conditions for growing shade-loving plants.

It may be made of wooden slats, or it may be similar to a greenhouse but with a covering of shade cloth.

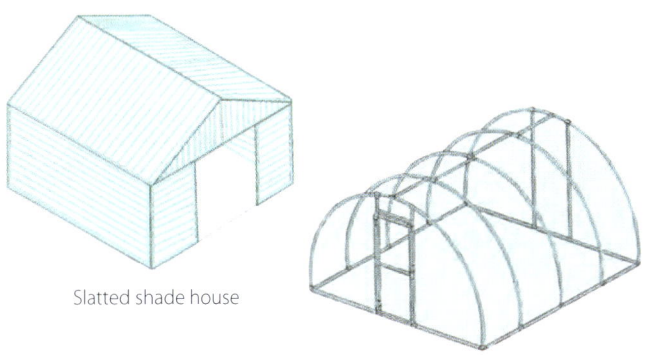

Slatted shade house

Polytunnel frame with shade cloth covering

## shearing = hedging

## sheet composting *see* page 122

## shrub

A shrub is a small- to medium-sized perennial plant with several woody stems rising from ground level. It has no distinct trunk and new growth is woody.

Shrubs can be evergreen or deciduous.

Rhododendrons and azaleas are shrubs.

Woody stems arise from ground level. There is no trunk.

### sheet composting

Sheet composting, also called lasagne composting, is a method of layering materials directly on the ground to improve the soil in a garden bed, to create a new garden bed, or to change a lawn to a garden.

The compost does not need to be turned, and after several months the layers rot down into a fertile soil that is ready to be planted out.

Methods of layering differ, but a basic plan is to first spread overlapping cardboard and newspaper on the ground to suppress weeds. Organic 'brown' layers of leaves, straw and coir are then alternated with nitrogen-rich 'green' layers of seedless grass clippings, coffee grounds and manure. These two layers can be repeated to achieve the required depth.

Finally, there is a compost layer that is topped with mulch. Each layer should be thoroughly watered. The bed will settle as the layers break down.

**SHEET COMPOSTING**

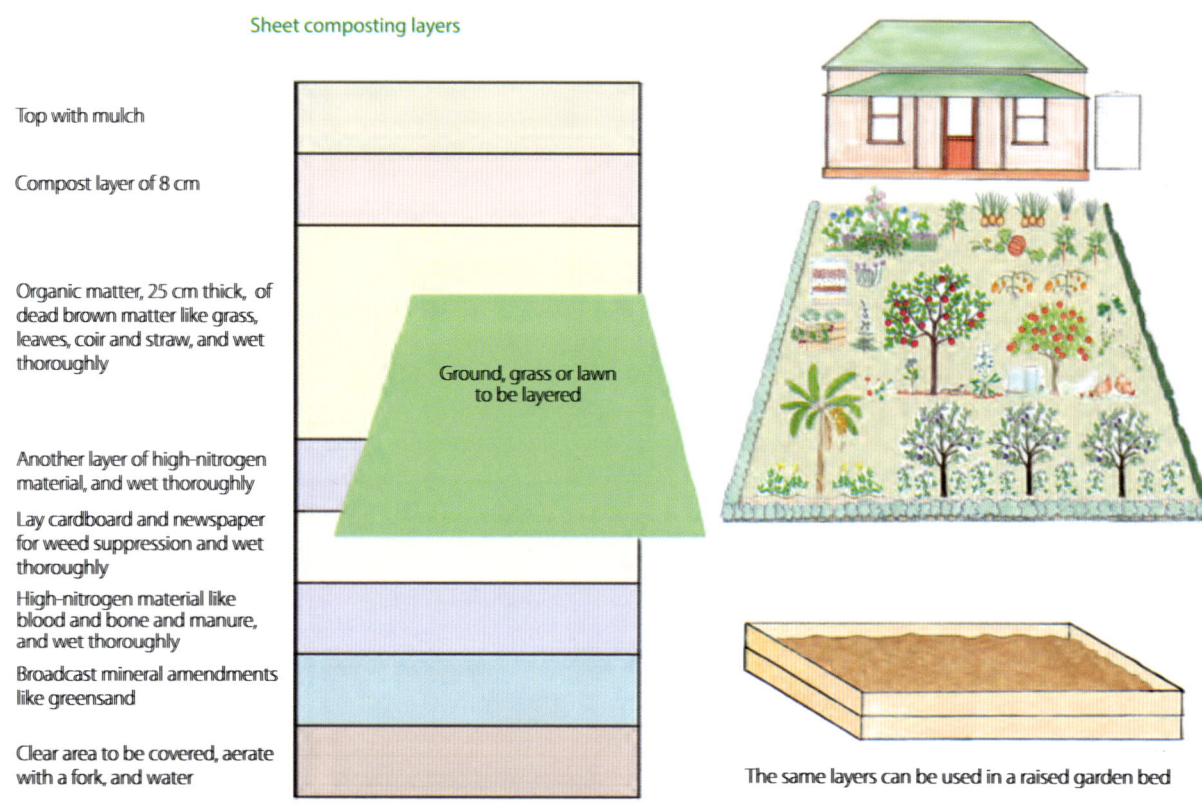

Sheet composting layers

Top with mulch

Compost layer of 8 cm

Organic matter, 25 cm thick, of dead brown matter like grass, leaves, coir and straw, and wet thoroughly

Ground, grass or lawn to be layered

Another layer of high-nitrogen material, and wet thoroughly

Lay cardboard and newspaper for weed suppression and wet thoroughly

High-nitrogen material like blood and bone and manure, and wet thoroughly

Broadcast mineral amendments like greensand

Clear area to be covered, aerate with a fork, and water

The same layers can be used in a raised garden bed

After *Modern Farmer* , 'Sheet Mulching' https://modernfarmer.com/2016/05/sheet-mulching/

### side grafting

In side grafting, a slanting angular cut is made on one side of a branch that is intended to be the rootstock. A slanting cut is also made on both sides of the scion wood. The cambium layers are matched, and the graft is tightly bound with tape.

After the union has healed, the rootstock is cut off just above the scion.

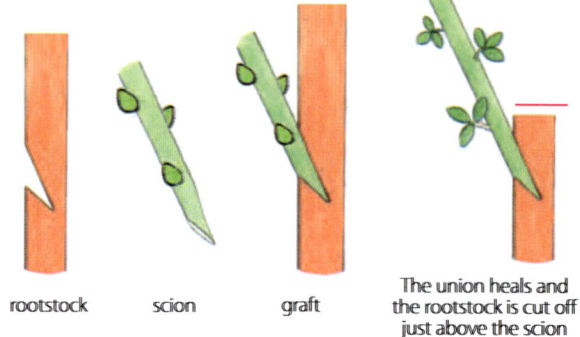

rootstock    scion    graft

The union heals and the rootstock is cut off just above the scion

## silt

Silt is a dust-like sediment made of rock worn away by water and ice. It is transported by rivers and wind. As a soil, it is fertile, light and moisture-retentive. Sand particles range in size from 0.05–2.0 mm, silt from 0.002–0.5 mm and clay < 0.002 mm.

## silty soil

Silty soil is usually more fertile than other types of soil and is light and moisture-retentive. A wide range of plants grow well in silty soil.

## simple layering

Simple layering is a method of propagation. A low-growing flexible branchlet is bent down, pegged in place, and covered with soil, leaving about 20–30 cm of the tip above ground.

The branchlet should root at the bend. The branchlet can also be wounded on the lower side of the bend to encourage rooting.

Simple layering can be used to propagate azaleas and rhododendrons

## simples

In the Middle Ages, simples were medicinal remedies using a single herb for the treatment of disease.

## slip

A slip is part of a plant that is cut or broken off for propagation. The slip will grow its own roots and foliage and will be a clone of the parent.

Many plants grow from slips, including those that are sterile and hybrids that will not grow true to type from seed. Sweet potatoes (*Ipomoea batatas*) can be grown from slips that form on tubers. A slip may also be used as a scion in grafting. It can also refer to a softwood cutting, especially of roses.

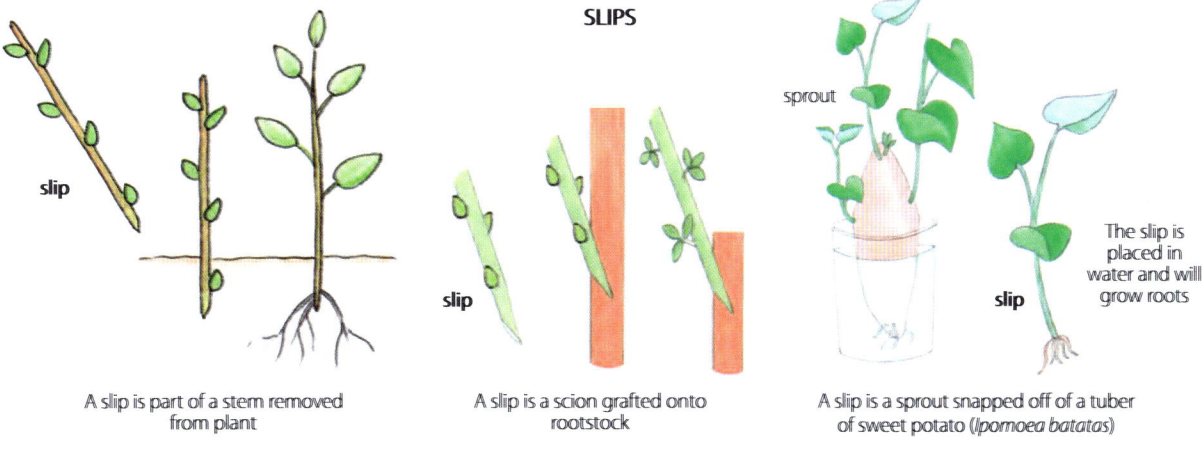

**SLIPS**

slip

slip

sprout

slip

The slip is placed in water and will grow roots

A slip is part of a stem removed from plant

A slip is a scion grafted onto rootstock

A slip is a sprout snapped off of a tuber of sweet potato (*Ipomoea batatas*)

## slipping

Slipping refers to bark that is easily peeled back rather than breaking or tearing. This occurs at the time of the year when the plant is in active growth.

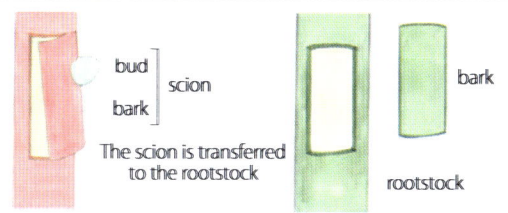

bud
scion
bark

The scion is transferred to the rootstock

bark

rootstock

Bark is easily peeled back and removed

### slow release fertilisers

Slow release fertilisers gradually release small amounts of nutrients into the soil. They include granules or pellets, and controlled release fertilisers that have coated pellets. Aside from the gradual release of nutrients, their main advantage is that they remain in the soil unaffected by heavy rain and are not washed away into the environment like water-soluble fertilisers.

### slugs and snails *see* pests and diseases of gardens

### softwood cuttings

Softwood cuttings are taken from young fast-growing stem tips in spring.

A length about 10 cm long is cut across at a node or a heel is taken by gently pulling off a side shoot from last year's stem. Lower leaves are removed, and larger leaves may be cut to half their length to minimise transpiration.

The cut end is dipped in rooting hormone and inserted into a potting mix. It needs to be kept in a moist, light environment.

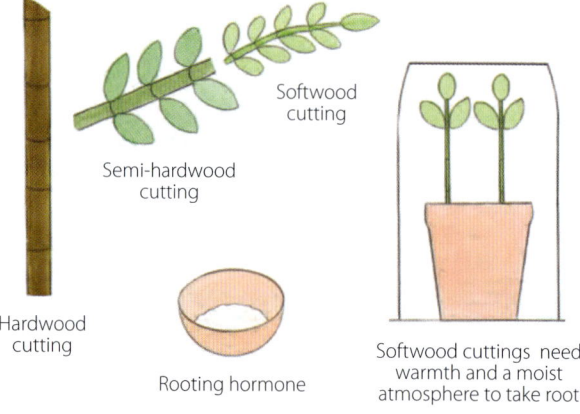

### soil

Soil forms from minerals in eroded rock, water and decayed organic matter.

Soil is made up of about 45% minerals, 5% organic matter, 20–30% water, and 20–30% air. The soil is alive and ever-changing. In its underground food web, nutrients are constantly broken down, absorbed, and recycled by soil organisms.

The main mineral components in soil are sand, silt and clay. Sand particles are the largest and clay the smallest, with silt in between. The relative proportions give soil its texture.

The amount of space between the particles influences how easily water moves through the soil and how much water the soil will hold. Heavy soil, with too much clay in proportion to silt and sand, holds water but gives it up to plants very slowly. Light soils with a high sand content are well-aerated but do not hold much water and are low in nutrients. The ideal mix, loam, is mostly considered to be about equal parts sand, silt and clay. The ideal topsoil for gardening would be loam.

The soil layer suitable for cultivation makes up about 7.5% of the Earth's surface.

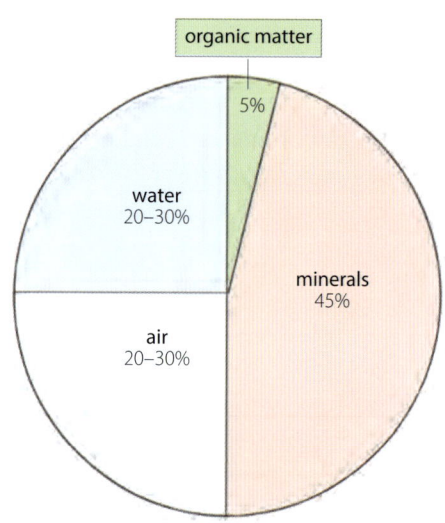

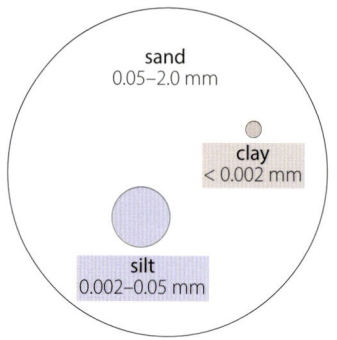

**Sand** is small, loose grains of disintegrated rock.
It is well-aerated and free-draining but has no capacity to hold water or nutrients.

**Silt** is rock worn away by water and ice, that is transported by rivers and wind.
It is fertile, light and moisture-retentive.

sand

silt

loam

clay

**Clay** particles are shaped like flakes. It retains nutrients but in winter it is wet and sticky and in summer it bakes hard and dry.

**Loam** is a mixture of clay, sand and silt. It is fertile, well-drained and easy to work.

## soil aggregates

Soil aggregates are units of soil consisting of clay, silt, sand and organic matter.

Sand and silt particles are held together by clay, organic matter like roots, and threads of fungal hyphae. Most soil bacteria live in these clusters.

Space between the particles in the aggregates provides pores for the exchange of gases with the air and spaces for water retention. Peds are natural soil aggregates, and crumbs and clods are aggregates formed when soil is tilled. Tilling destroys the natural peds, which are best left undisturbed.

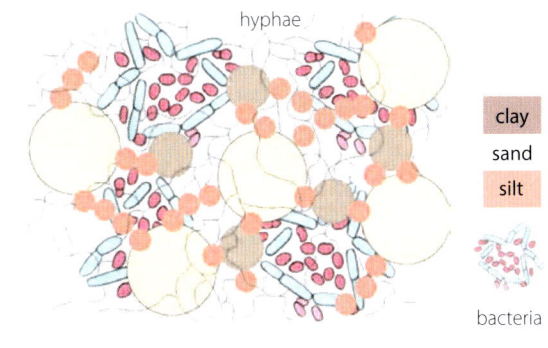

hyphae

clay

sand

silt

bacteria

A soil aggregate

## soil biota *see* soil food web

## soil carbon

Carbon is stored mainly in plants and soil. Plants draw carbon from the air through photosynthesis and store it in their leaves, stems and roots. When they die and rot down, the carbon is stored in the soil. Carbon that is not used by the living plant is exuded through the roots to feed mycorrhiza and other beneficial soil organisms.

Soil loses stored carbon to the atmosphere when a landscape is cleared or tilled. Chemical fertilisers work adversely by inhibiting microbial and fungal interactions that store carbon in the soil. Peat moss, and peat, which stores one third of the world's carbon, are best left in the ground untouched.

Many garden practices can improve the level of soil carbon. Levels are increased by using cover crops with an extensive root system and growing plants like pigeon peas for chop and drop. Minimising tillage protects soil structure and prevents the release of soil carbon into the atmosphere. Moving animals so that plants are not overgrazed allows the soil to continue to benefit from both carbon storage provided by the green ground cover and animal waste.

KEEPING CARBON IN THE SOIL

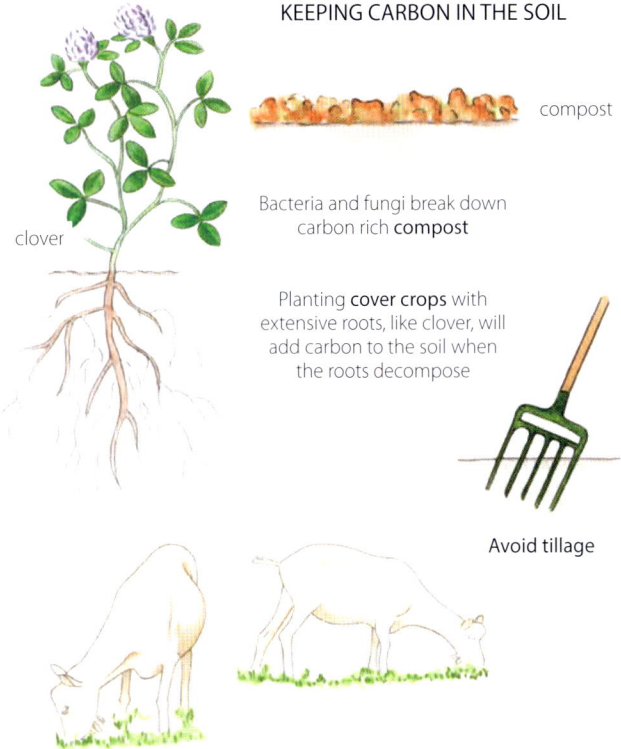

compost

Bacteria and fungi break down carbon rich **compost**

clover

Planting **cover crops** with extensive roots, like clover, will add carbon to the soil when the roots decompose

Avoid tillage

Good **grass cover** adds carbon to the soil

Moving animals in a timely way means soil continues to benefit from the grass and from animal waste

## soil erosion

Soil erosion is the process by which the land surface is worn down and the materials transported away by wind and water to another site. It includes the removal of topsoil after it has been exposed by clearing vegetation, tillage, mining and other means.

## soil food web

A soil food web is a community of organisms that live all or part of their lives in soil.

They have intricate relationships with each other, and with the organic matter and minerals in the soil. The food web makes for healthy soil if its cycle of decomposition and recycling of nutrients is left undisturbed.

**SOIL FOOD WEB**

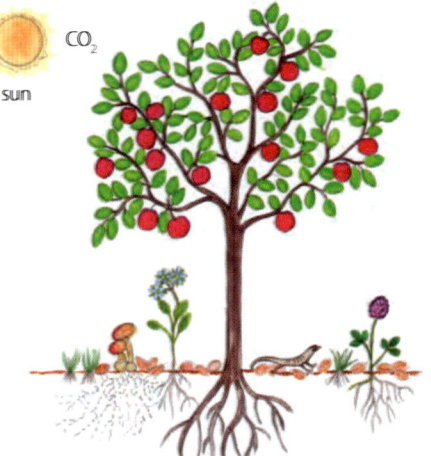

### Photosynthesis

The soil food web begins with energy from the sun that triggers photosynthesis in plants.

Photosynthesis fixes carbon dioxide from the atmosphere in the leaves. The plant uses this to produce oxygen and to make organic compounds like starches and sugars.

### Organic matter

Organic matter is the remains of plants, animals and microbes.

It is recycled by soil organisms and is eventually turned into humus that can be broken down into nutrients that are absorbed by plant roots.

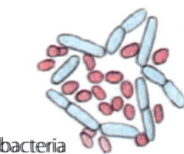

bacteria

hyphae

### Bacteria

Together with fungi, bacteria are a primary decomposer of organic matter. They make up the largest number and weight of any soil microorganism.

### Fungi

Together with bacteria, fungi are a primary decomposer of organic matter.

They secrete substances that break down the organic matter so that the hyphae can absorb it. The network of hyphae (a mycelium) transports these nutrients to plant roots through vast areas of soil.

As thread-like hyphae grow they entangle and enmesh particles into larger clumps making air spaces, giving the soil a better, more open texture.

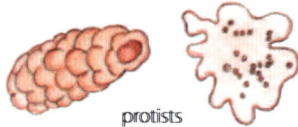

protists

### Protists

Protists are predators that are the main consumers of bacteria and fungi.

Protists 'recycle' bacteria and fungi by eating them. At the same time, they control the size of these communities.

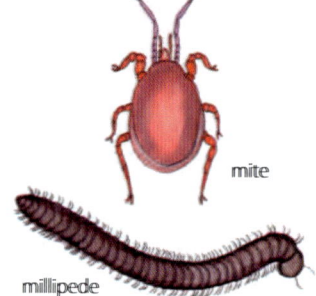

mite

millipede

### Arthropods

Include mites and millipedes that feed on organic matter, other arthropods, fungi and worms. As they move around, they turn and aerate the soil and leave nutrients in their droppings.

nematodes

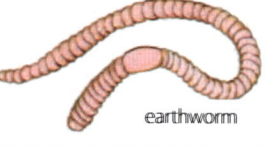

earthworm

### Nematodes

Nematodes are minute round worms that feed on fungi, bacteria, protists and other soil organisms.

### Earthworms

Earthworms have bodies that are divided into segments. They eat organic matter and soil as they tunnel, and leave fertile castings as they go. Their tunnelling increases the amount of water the soil can hold.

## soil pH

Soil pH is a measure of the acidity or alkalinity of the soil. The pH scale ranges from 0 to 14. Generally, a soil pH between 6.0 and 7.5 is acceptable to most plants.

Soil pH determines nutrient availability to plants. Some plants, like blueberries, prefer a soil pH of around 4 or 5. Others, like lindens and elms, can tolerate some degree of alkalinity.

Litmus paper and various instruments are available for measuring the pH of garden soils. Litmus paper comes in strips that are dipped for about 60 seconds into a mix of equal parts soil and distilled water. The colour of the strip is then compared with the colour scale key included in the kit to find out the pH.

*see also* **acidic soil, alkaline soil**

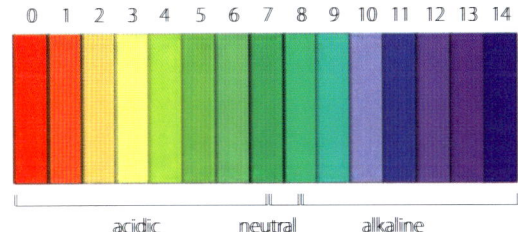

Soil pH classification

| | |
|---|---|
| over 7.5 | alkaline |
| 6.5–7.5 | neutral |
| less than 6.5 | acidic |

The colour scale key for measuring pH using litmus paper

## soil profile

A soil profile is a vertical cross-section of a soil showing the distinct horizontal layers called horizons.

Different soils will have different profiles. Most soils have three main horizons: the topsoil, the subsoil and the parent material that is located just above the bedrock.

**SOIL PROFILE**

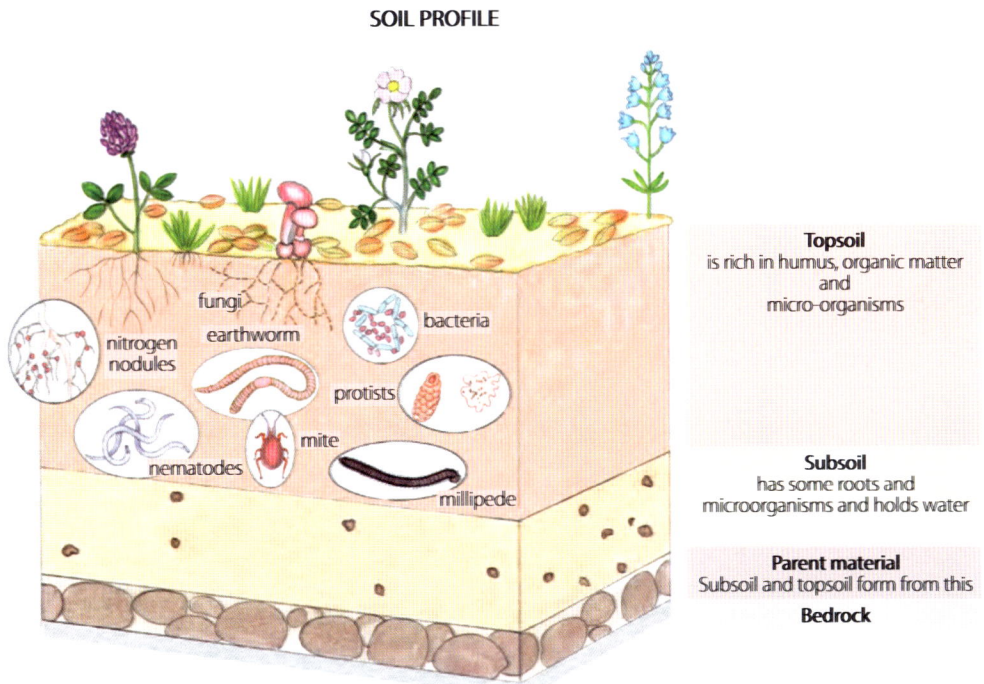

**Topsoil**
is rich in humus, organic matter and micro-organisms

**Subsoil**
has some roots and microorganisms and holds water

**Parent material**
Subsoil and topsoil form from this
**Bedrock**

## soil structure

Soil structure refers to how particles of sand, clay and silt group together naturally to form aggregates (peds). These aggregates occur in different patterns and vary in size. Poor soil structure has few aggregates.

A well-structured soil allows air and water to move through the soil for healthy plant growth.

**soil texture**

Soil texture refers to the proportion of sand, silt and clay particles that make up the mineral component of soil.

**soil wettability**

Soils either absorb water readily (are hydrophilic) or repel water (are hydrophobic). If soil is hydrophobic it is unable to absorb moisture. Water cannot penetrate to the roots, and plants are unable to absorb the nutrients they need.

Soil becomes hydrophobic when a waxy coating, caused by fungi or decomposing organic plant matter, covers the soil particles. Usually, the coating is broken down by soil microbes. Tilling affects soil wettability. It destroys soil porosity, so that water is unable to pass through.

Wetting agents are a remedy for hydrophobic soils.

**solarisation**

Solarisation uses heat from the sun to control pests, soil-borne diseases, and weeds in the soil.

The ground is watered deeply, then covered, usually with a transparent polyethylene cover that is sealed around the edges. It works by trapping the sun's energy and superheating the soil. The process needs daily full sun over a period of about 4 to 8 weeks.

It can be used on any scale and is effective in a backyard garden as well as a market garden. It also works on raised garden beds.

Clear plastic allows heat from the sun to get into the soil, white or black plastic blocks it

Edges are buried in a trench so that the plastic is held as tightly as possible against the soil

**sooty mould fungus** *see* **honeydew**

**sour soil** = acidic soil

**sowing seeds**

Seeds are sown directly in the garden where they grow, or they are raised under cover and planted out later as seedlings.

Seeds that are best sown directly into the garden include root vegetables and legumes. Almost all seeds are sown at a depth that is twice their size. Some, like lettuce and poppies, need light to germinate and are covered with a sparse layer of soil. There are special seed raising mixes for growing seeds indoors. Soaking seeds before planting can hasten germination.

**species,** *abbr.* **sp.,** *pl.* **species,** *abbr.* **spp.**

Species refers to the basic unit of plant classification. It is a group of individuals that can interbreed and produce fertile offspring.

A species has two names: the first is the genus name, and the second is the species name, as *Prunus avium* (sweet cherry).

*see also* **genus, plant kingdom**

## sphagnum moss

Sphagnum moss is a soft, pliable moss that is dried for use in potting mixes, garden soils, seed starting mediums and basket liners.

It has a neutral pH and is highly water-retentive. Long-fibred sphagnum is left in its natural form, and milled sphagnum has been finely chopped.

Sphagnum is a living plant harvested from wetlands on top of peat. When it dies, it does not decay readily and accumulates over hundreds, or thousands, of years eventually becoming peat.

Sphagnum moss filters water and prevents rain from evaporating. These mosses are part of a wetland plant community.

There are over 350 species of *Sphagnum*.

*Sphagnum squarrosum*

Sphagnum moss basket liner.
Coir is now widely used instead.

Sphagnum moss potting mix.
Coir can be used as a replacement.

## splice grafting

In splice grafting, a simple slanting cut of the same length is made on both the scion and rootstock. The two are lined up and bound together.

Splice grafting is usually done to herbaceous plants, like cucumbers, that are grafted onto disease-resistant rootstock.

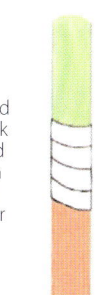

scion      rootstock

Scion and rootstock are lined up then bound together

## split vein propagation *see* leaf cuttings

## spore

A spore is the reproductive unit that takes the place of seeds in plants like ferns and mosses.

## sport

A sport is a plant or part of a plant, such as a leaf or flower, that differs from the rest of the species. It is usually transient, but it may be propagated vegetatively to form a new cultivar. It results from a genetic mutation.

Sports affect flower colour but can also affect growth and blooming. The hybrid tea rose 'Peace' released in 1945, is famous for its very fine sports including 'Climbing Peace'. Sports can also be fruits. The nectarine is a sport of the peach.

## sprinklers *see* irrigation

## sprouts

Sprouts are very young plants that are harvested just a few days after they germinate. The seed, root and stem are eaten.

To grow sprouts, seeds are soaked for several hours and then exposed to the right combination of temperature and moisture so that they germinate. They can be grown in a commercial sprouter, a covered jar or a seed tray. Seeds grown for sprouts include mung beans, alfalfa and snow peas.

## spur

A spur is a short shoot of buds that forms on fruit trees.

Trees that have spurs include peaches, plums, apricots and cherries, as well as apples, pears, almonds and walnuts.

Spurs form differently according to the type of tree, and this requires attention when pruning.

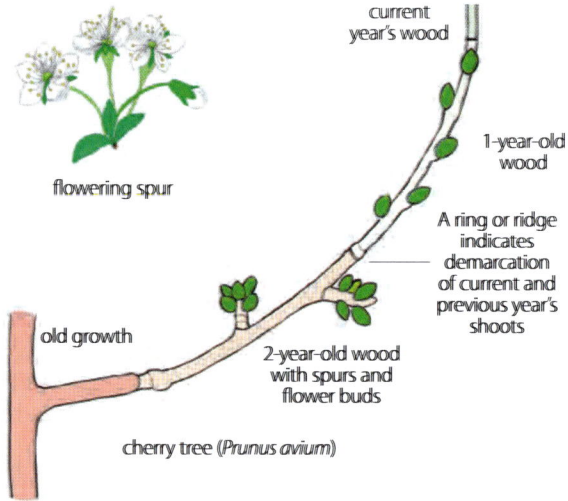

flowering spur

current year's wood

1-year-old wood

A ring or ridge indicates demarcation of current and previous year's shoots

old growth

2-year-old wood with spurs and flower buds

cherry tree (*Prunus avium*)

## square foot gardens

A square foot garden is a raised garden bed divided into 30 cm squares.

The bed is commonly 1 m wide by 1–2 m long, with rich soil that is at least 20 cm deep. Plantings depend on the size of the plant. For example, in a 30 cm square you may fit 16 carrots, 9 lettuces or one eggplant.

This small but intensive garden incorporates methods from organic gardening. These include succession planting, composting, companion planting and mulching. Plant species are usually grown again in a different square.

Climbing plants, like pole beans and peas, and surface runners, like cucumbers, can be grown on a lattice. Some vegetables that require more space to grow, like artichokes and asparagus, are best avoided in a square foot garden.

trellis facing the sun for climbing plants

| cucumber | climbing peas | tomato | pole beans |
| peppers | broccoli | eggplant | Brussels sprouts |
| spinach | chamomile | basil | mint |
| carrots | spring onions | thyme | lettuce |

An example of a 1 m × 1 m square foot garden

**stamen** *see* **flower**

**standard** *see* **topiary**

## stem

A stem is the above-ground part of the plant that usually grows upwards, with roots at the base. Some stems, however, like stolons, grow horizontally, and some, like rhizomes and tubers, grow underground. These have roots growing from the nodes on the stem.

Stems may be herbaceous or woody. An above-ground stem supports leaves, flowers and fruit. It also transports nutrients. Woody stems may be hardwood, semi-hardwood or softwood depending on their age.

All stems, whether above- or below-ground, have nodes from which leaves, shoots and aerial stems grow. The space between the nodes is the internode.

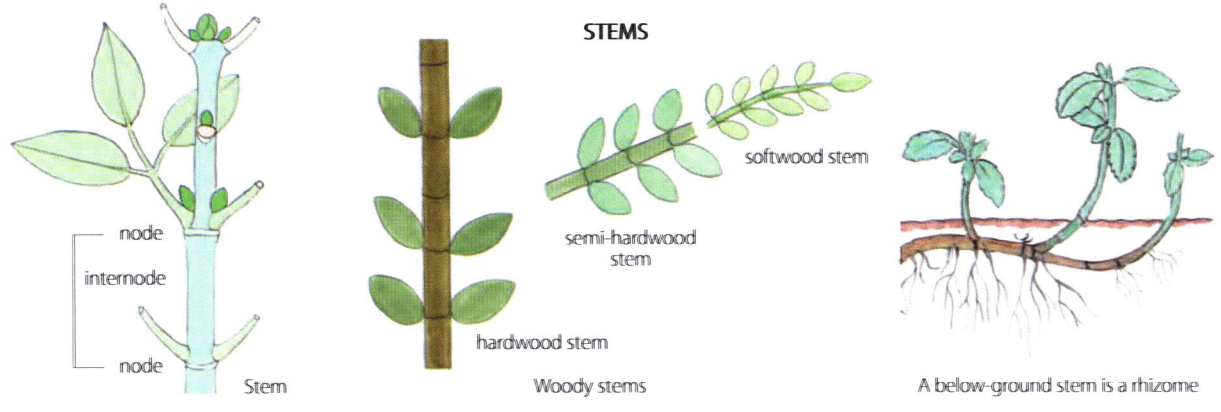

**STEMS**

node
internode
node
Stem

hardwood stem
Woody stems

softwood stem
semi-hardwood stem

A below-ground stem is a rhizome

## stem cuttings

There are four types of stem cutting used for propagating plants. Both evergreen and deciduous woody plants can be propagated by softwood, semi-hardwood and hardwood cuttings. Herbaceous cuttings are taken from the new, soft green growth of non-woody plants.

Softwood cuttings are taken from new, green, non-woody growth. Semi-hardwood cuttings are taken from wood between the softwood and hardwood stages; the wood is firm but still flexible and will snap easily. Hardwood cuttings are fully mature wood that will bend slightly, and are taken from deciduous trees during dormancy. A stem or small branch may have hardwood, semi-hardwood and softwood along its length.

Cuttings can be taken from the tip of a shoot. The lower leaves are removed and larger leaves may be cut to half their length to minimise transpiration. Cuttings with nodes, from which new growth starts, are typically placed in the soil with the lower nodes covered and the upper nodes exposed. Heel and mallet cuttings have a small part of the parent woody branch attached to the base with cambium cells exposed that produce new roots.

Applying a root-stimulating hormone to the base that will be in the soil increases the success rate of stem cuttings.

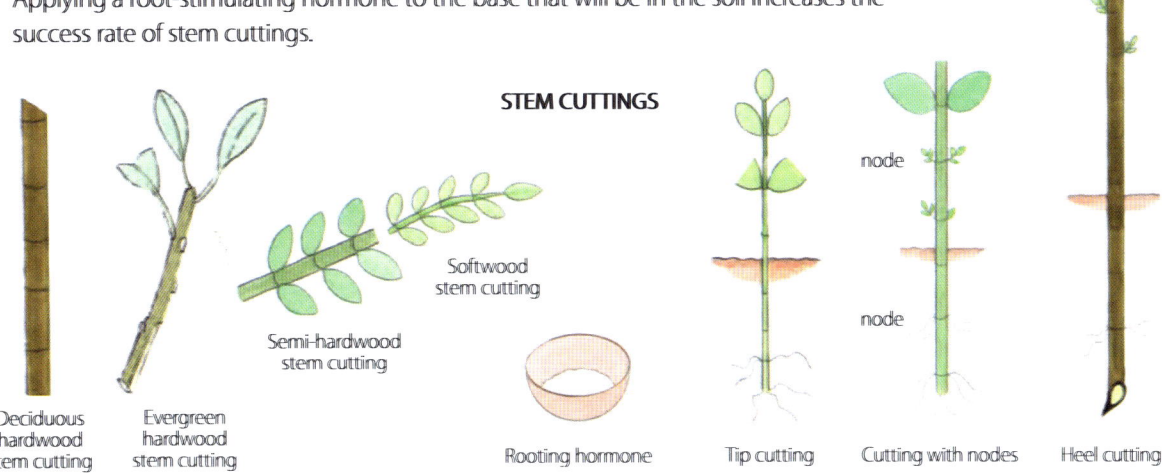

**STEM CUTTINGS**

Deciduous hardwood stem cutting

Evergreen hardwood stem cutting

Semi-hardwood stem cutting

Softwood stem cutting

Rooting hormone

Tip cutting

node

node

Cutting with nodes

Heel cutting

## stem tuber

A stem tuber is an enlarged part of an underground stem (a stolon or a rhizome).

The stem has tiny scale leaves with buds that grow on its surface. Each of these buds can form a new plant that is a clone of the parent. Examples are potatoes (*Solanum tuberosum*), with tubers that form on stolons, and Jerusalem artichokes (*Helianthus tuberosus*), with tubers that form on rhizomes.

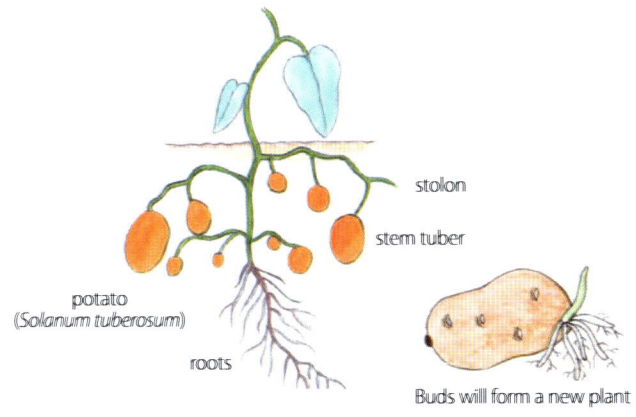

stolon

stem tuber

potato
(*Solanum tuberosum*)

roots

Buds will form a new plant

131

**stigma** *see* flower

**stock** = rootstock

**stolon**

A stolon is a horizontal stem that arises from near the base of the parent plant. The stem rests on the ground, and new plantlets root at the nodes or at the tip. Plantlets are capable of independent growth when a stolon is divided or dies away.

Strawberries produce stolons. Some plants, like mints, produce stolons just below the soil surface. Unlike rhizomes, stolons do not store nutrients. Stolons are sometimes called runners or offsets.

parent plant

stolons

strawberry (*Fragaria*)

Stolon divided for propagation

**stone fruit** *see* drupe

**stool layering** = mound layering

**stratification**

Stratification is a natural process that brings about internal changes in some seeds before they can germinate. It is a survival mechanism so that seeds do not germinate prematurely.

Some seeds require a period of warmth and moisture before germination. Others require cold, moist conditions. Stratification can be done artificially prior to sowing when the requirements of particular seeds are understood.

**straw bale garden**

Straw bale gardens were developed by American Joel Karsten in the 1960s. It can be used on any soil but needs to be watered regularly. It lasts for only one or two years, then the finished bale is used as compost.

Choose straw, not hay. Straw bales are made from the stalks of cereal grain crops. Hay is used to feed livestock and contains seeds.

Before planting, condition the bale over several days by adding water and nutrients.

Straw bale size
40 × 50 × 100 cm

Suitable plants include salad vegetables, strawberries, tomatoes, peppers and eggplant

**style** *see* flower

**subshrub**

A subshrub is a small shrub with a woody base and herbaceous new growth. A shrub has woody new growth.

Russian sage (*Perovskia atriplicifolia*) is a deciduous subshrub, and lavender (*Lavandula*) and thyme (*Thymus*) are evergreen subshrubs.

new growth is herbaceous

lavender (*Lavandula*)

stems woody at the base

## subsoil

Subsoil is a nutrient and water reserve below the topsoil. It has less organic matter and microorganisms than the topsoil, and fewer roots.

## subspecies, *abbr.* subsp., *pl.* subspecies, *abbr.* subssp.

Subspecies refers to the unit of plant classification below species and above variety, such as the wild cherry, *Prunus avium* subsp. *avium*.

*see also* plant kingdom, variety

## succession planting

Succession planting is a method of sowing a crop at 7 to 21-day intervals so that vegetables can be harvested throughout the season instead of all at once. Planting seeds of the same crop with different maturity dates is another method of succession planting.

Staggered plantings work well for beets, carrots, onions and lettuces that are harvested only once.

1 week    4 weeks    7 weeks    10 weeks

Succession planting of spinach.
Spinach takes 7 weeks to mature.
Sow seeds every 2–3 weeks.

## succulents

Succulents have fleshy, water-storing stems and leaves that allow them to survive in unforgiving conditions like dry landscapes, rocky terrain and nutrient-poor soil. They occur in several botanical families and are native to different regions across the world.

There are more than 10 000 varieties of succulents, with options for indoor plants, rock gardens, crevices, xeriscaping and borders, or as a contrast with other plants in the garden. They are attractive for their easy care and multitude of forms. Lithops looks like a stone. Yucca has impressive blooms with some having flowering stems to 3 m high, while string of pearls is named for its rounded leaves.

'Succulents' is also a common term for a member of the cactus family (Cactaceae). Though there are similarities, they are not considered by many to be a member of the succulents group of plants. Cacti seldom have leaves and are distinguished from other succulents by having tiny cushion-like areoles from which branches, flowers and spines arise.

SUCCULENTS

Zebra plant
(*Haworthia fasciata*)
is a delicate small plant with slender
pointed leaves banded with white.
It is popular as an indoor plant.

Blue chalk sticks
(*Senecio serpens*)
is a spreading ground cover or
contrasting border plant that is also
well-suited to rockeries

String of pearls
(*Senecio rowleyanus*)
has small pearl-like leaves along
a dangling stem, making it ideal
for hanging baskets

## sucker

A sucker is a shoot that develops from an underground bud on the root system of a plant.

Usually it is removed if it is detrimental to the parent plant. Suckers can be used to propagate a clone of the parent plant.

Sucker shoots from the rootstock of a graft, if allowed to grow, will have the characteristics of the rootstock tree, not the cultivar.

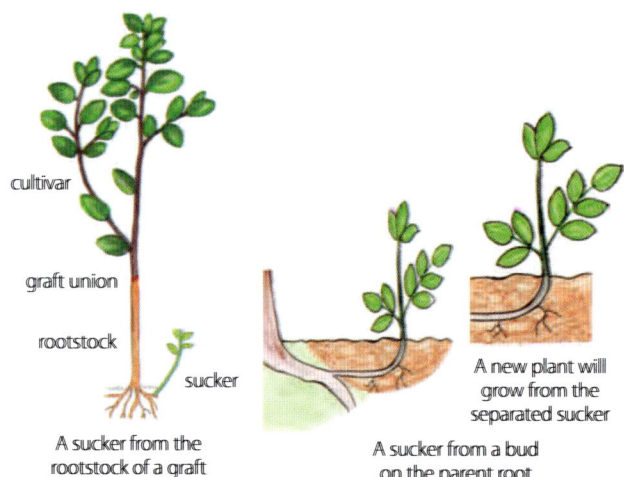

cultivar

graft union

rootstock

sucker

A sucker from the rootstock of a graft

A new plant will grow from the separated sucker

A sucker from a bud on the parent root

## sulfur *see* plant nutrients

## swale

A swale is a ditch dug around a contour so that it is level along its length. The aim of a swale is to capture rainwater.

The base of the swale is flattened so that water spreads evenly along its length and sinks into the soil. Usually, the earth that is dug out is placed on the downhill side to make a berm. The berm is stabilised by first planting fast-growing trees that are later gradually replaced with productive trees. The upside of the swale is also planted, for example, with leguminous plants.

The swale is dug on a slope of no more than 15**%**. Generally, it is twice as wide as it is deep. A swale 1 m wide would be 0.5 m deep.

The water table should be well below the surface so that the water captured in the swale can soak away readily.

Berm and swale systems are ideal for growing trees

**SWALE**

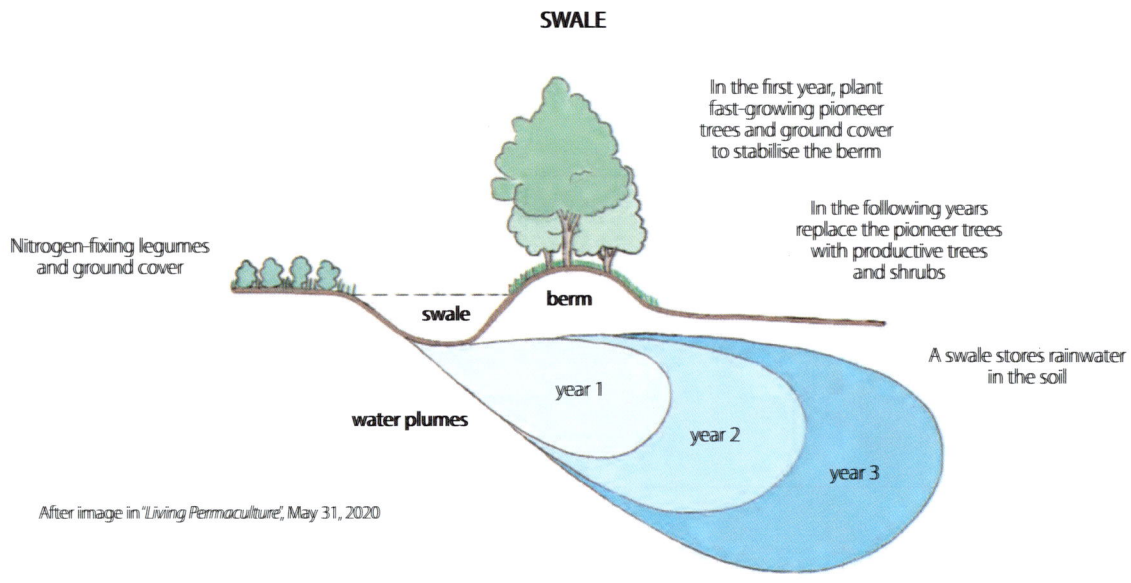

In the first year, plant fast-growing pioneer trees and ground cover to stabilise the berm

In the following years replace the pioneer trees with productive trees and shrubs

Nitrogen-fixing legumes and ground cover

swale

berm

A swale stores rainwater in the soil

year 1

water plumes

year 2

year 3

*After image in 'Living Permaculture', May 31, 2020*

**sweet soil** = alkaline soil

134

## T-budding

T-budding is a form of grafting. An incision is made in the bark of the rootstock in the shape of the letter 'T'. This exposes the cambium. A piece of bark, shaped like a shield, with a single bud, is removed from the plant to be propagated. This is the scion.

The cambium of the scion is lined up with the cambium of the rootstock when inserting it in the T-shaped cut. It is then secured firmly with tape. The bud, with its desired characteristics, will grow to form the top of the tree.

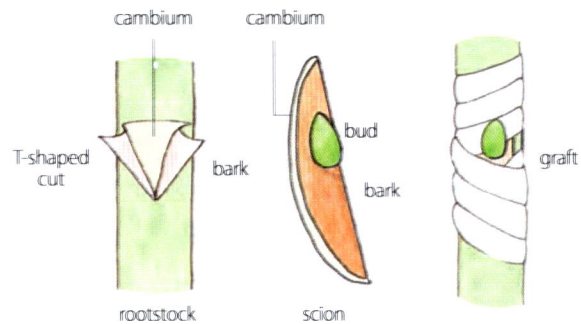

## tender

Tender plants are damaged by frost and cannot survive harsh conditions.

## tendril

A tendril is a slender, stem-like structure used by a plant to twine around a support. It is a modified stem, leaf, petiole or other structure.

Plants that have tendrils include peas (*Pisum*), cucumbers (*Cucumis*) and grapevines (*Vitis*).

Tendrils of peas are modified leaflets

A pumpkin tendril is a modified stem

## terminal bud *see* bud

## terrarium

A terrarium is a closed or open system for creating a miniature garden indoors.

Closed systems mimic the natural world. Moisture from the soil and leaves condenses on the walls of a glass container. The condensation drips down over the plants and, like rain, provides water. The clear glass allows the leaves to photosynthesise. The container is sealable but can be opened for cleaning and plant care.

Generally, the most suitable plants for a sealed terrarium are those that thrive in humid environments, like orchids, ferns and mosses.

An open terrarium is used for growing plants, like succulents, that prefer dry conditions.

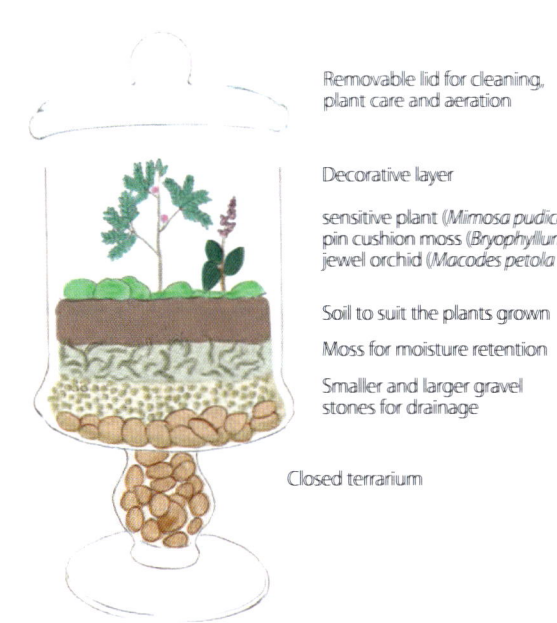

Removable lid for cleaning, plant care and aeration

Decorative layer

sensitive plant (*Mimosa pudica*)
pin cushion moss (*Bryophyllum*)
jewel orchid (*Macodes petola*)

Soil to suit the plants grown

Moss for moisture retention

Smaller and larger gravel stones for drainage

Closed terrarium

## thatch

Thatch is a layer of dead plant material that builds up on the soil surface under turf.

Thatch that is more than 25 mm thick needs to be removed because it restricts water and nutrients from reaching a lawn's roots. If the thatch is not too thick, it can be removed with a thatching rake. Specialised dethatching machines are also available for larger lawns. Scalping is done by setting the mower as low to the ground as possible and using a grass catcher.

## thinning

Thinning a perennial shrub is done by removing some stems to or near ground level before it reaches a mature size. Thinning is also done to the canopy of a tree.

The removal of weak or thin stems makes the plant more sturdy. Thinning reduces the risk of disease by increasing air circulation and sunlight levels. It is a way of keeping the shape of the plant more natural and pleasing without significantly reducing its size.

Growth is dense, allowing little air flow or sunlight

Thinning opens up the structure of the plant

## thinning out

If seedlings are crowded together, some need to be removed. This allows space for air and for the remaining plants to grow without competition. Unwanted seedlings are snipped with scissors at soil level, leaving the remaining plants to develop without their roots being disturbed.

## Three Sisters Garden

The Three Sisters Garden is an ancient style of companion planting practised by North American Indians using corn, beans and squash or pumpkin.

Corn provided support for climbing beans, and the large squash leaves provided a living mulch and suppressed weeds. Beans added nitrogen to the soil. The combination enriches the soil and results in high yields. Some peoples added sunflowers to attract pollinators.

The gardens were grown on small, well-drained mounds of organic matter. No tilling was required. Across the continent, gardening styles varied according to climate. The circular garden of the Wampanoag people is an example.

Permaculture has adopted the Three Sisters Garden as an exemplary guild.

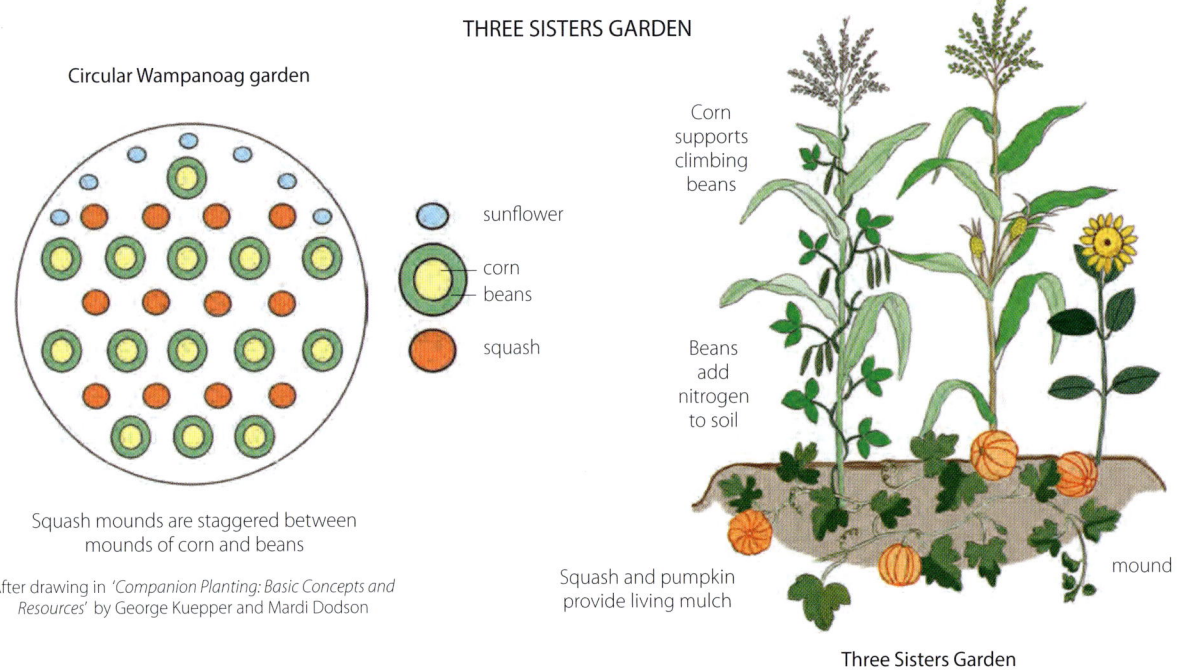

THREE SISTERS GARDEN

Circular Wampanoag garden

sunflower
corn
beans
squash

Squash mounds are staggered between mounds of corn and beans

After drawing in *'Companion Planting: Basic Concepts and Resources'* by George Kuepper and Mardi Dodson

Corn supports climbing beans

Beans add nitrogen to soil

Squash and pumpkin provide living mulch

mound

Three Sisters Garden

## three-bin composting

Three-bin composting is a hot composting method. Microorganisms that break down the compost create heat that decomposes the material.

Layers of chopped green and brown composting material are added to bin 1, in a ratio of 1 part green to 2 parts brown. New material should be placed only in bin 1. Once bin 1 is full and the material is heating, the contents of bin 1 are transferred to bin 2. Forking the material into bin 2 aerates it. Bin 1 can now be refilled with new material.

The composting material moved to bin 2 should now be covered and left alone to heat to 70° C, the optimal temperature for microorganisms to decompose the material. When bin 1 is full again, the contents of bin 2 are transferred to bin 3. The new material in bin 1 is placed in the now empty bin 2.

Composting material in bin 3 should be covered and allowed to cure for a few weeks while it continues to break down and cool. The compost should then be ready to place on your garden.

### THREE-BIN COMPOSTING

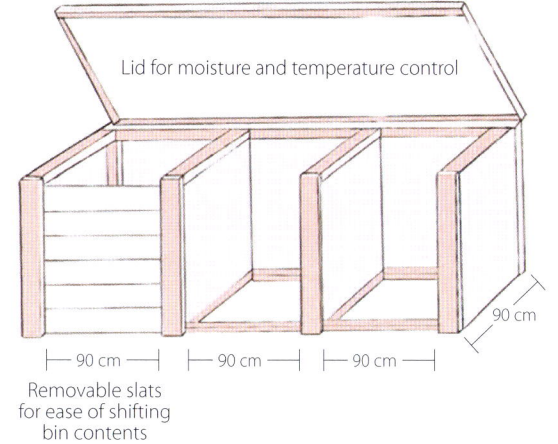

**Green material**
vegetable and fruit scraps
coffee grounds
tea bags
fresh grass clippings
chicken/livestock manure
green plant cuttings

**Brown material**
leaves
shredded straw, hay
sawdust
ashes from wood
shredded paper

Lid for moisture and temperature control

90 cm
90 cm | 90 cm | 90 cm

Removable slats for ease of shifting bin contents

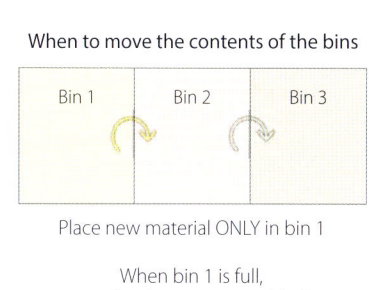

When to move the contents of the bins

| Bin 1 | Bin 2 | Bin 3 |

Place new material ONLY in bin 1

When bin 1 is full, transfer the contents to bin 2

The contents of bin 2 are transferred to bin 3 when bin 1 is again full of new material

## tillage

Tillage is the digging and turning over of soil, with hand tools or a machine, before planting. Tilling breaks down the natural soil structure that has spaces for air and water to move through. It exposes soil to loss of moisture, loss of carbon and to erosion. Tilling disturbs the network of life in the soil and buries weed seeds.

Minimal tillage practices leave the soil unturned, and some plant residues stay in the soil after harvest to protect its structure and retain moisture.

With zero tillage, soil is disturbed only where seeds and plants are planted.

## tiller

A tiller is a shoot arising at the base of the stem of a grass or sedge.

Tillers are separated from the parent plant for propagation. Once established, they can also produce an inflorescence and seeds.

Grass tiller

## tilth

Tilth is the condition of the soil as it relates to plant growth.

Favourable tilth implies good conditions for seed germination and ease of root penetration. Good tilth also facilitates other processes, such as water infiltration and aeration. It is usually equated with the presence of good soil crumb quality (aggregates) because stable aggregates promote these favourable processes.

Poor soil tilth is associated with compaction, animal trampling and tilling.

## tip layering

Tip layering is a method of propagation of plants, like cane berries, that prefer to root at the tip rather than along the stem.

The tip of a long, arching stem that easily reaches ground level is buried in the soil and pegged in place. Roots of new plants will eventually develop from the shoot tip.

Plants that can be tip layered include raspberries, jasmine, oleander and bougainvillea

## tip pruning

Tip pruning, also called pinching, is the removal of the soft growing tips of a new shoot. It is often done with a finger and thumb.

Each tip that is removed will have new shoots emerging beneath it. Tip pruning increases the number of shoots that will bear flowers next season and makes the plant more bushy.

tip removed

new flowers and lateral shoots form

buds in leaf axils

mint (*Mentha*)

**tissue culture** *see* **micropropagation**

**top grafting** = **topworking**

## topdressing

An even spreading of soil, a soil amendment, or fertiliser over the surface of the soil in garden beds, under trees, in pots or on lawns to increase soil and plant health.

## topiary

Topiary is an old art of training plants into geometric shapes, such as birds and animals.

Plants are trained and trimmed, sometimes over a wire frame, to get the desired shape.

A standard plant is trained to have a bare stem or trunk with a head of branches. Roses, fuchsias and many trees can be grown this way.

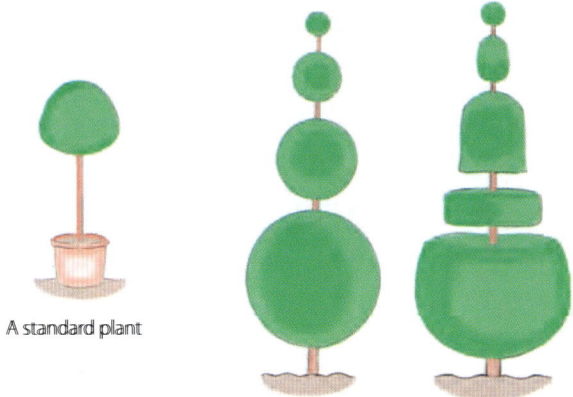

A standard plant

Any plant that is evergreen, has small foliage and regrows after pruning can be used for topiary. These include box species, camellias, evergreen azaleas, junipers and the Australian lilly pilly.

*see also* **cloud pruning**

## topping

Topping is the removal of the upper ends of the main branches of a tree. They are usually cut off at a uniform height.

Topping affects the tree's structure and appearance. The thin upright branches that grow from the cut branches are weak, prone to disease, and unsightly.

**TOPPING**

topped tree

new weak growth

## topsoil

Topsoil is the most fertile and biologically active layer of the soil. It is rich in humus, organic matter and microorganisms. Most plant roots are concentrated here.

## topworking

Topworking, also called top grafting, refers to grafting a new cultivar onto an established tree.

It is used to rejuvenate trees that are fruiting poorly, or to insert a cultivar that is disease-resistant and fruits prolifically. It avoids having to plant a new tree and hastens fruiting.

A healthy tree is cut back to a stump (rootstock). One or more scions from a new cultivar are grafted onto the rootstock. The tissues of the scions and the rootstock unite to form a new tree.

Bark grafting and wedge grafting are common methods used in topworking.

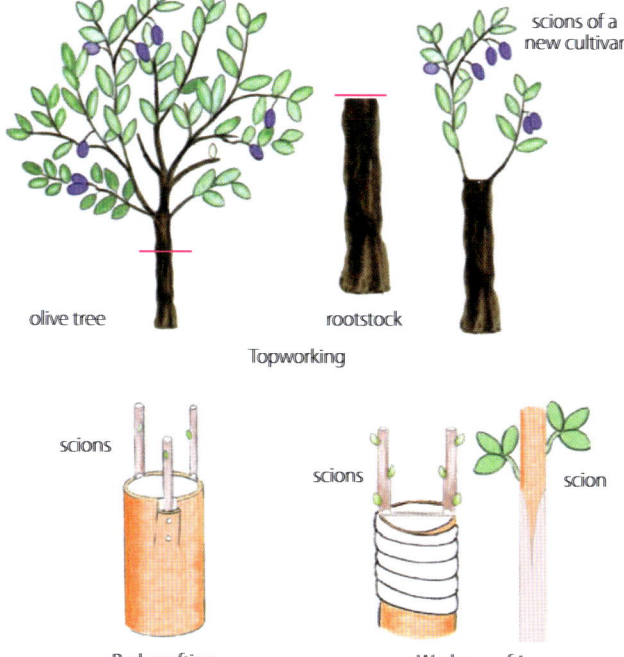

scions of a new cultivar

olive tree

rootstock

Topworking

scions

Bark grafting

scions

scion

Wedge grafting

## trace elements

Trace elements occur naturally in the soil in minute amounts. They include trace metals, heavy metals and micronutrients. Some, like boron, copper and manganese, are essential micronutrients for plants. Others, like arsenic, mercury and cadmium, in excessive quantities, can contaminate soil.

## trailers

Trailers have long stems that hang loosely or spread over the ground.

Trailers can be bold perennials like bougainvillea (*Bougainvillea*) or annuals like nasturtium (*Tropaeolum*). The Australian native laurel-leaf grevillea (*Grevillea laurifolia*) and its cultivars are trailing shrubs that are used as ground cover. Many trailers, like lobelia (*Lobelia erinus*), spill beautifully out of hanging baskets.

## transplanting

Transplanting is the movement of a tree, plant or seedling from one location to another.

This is done in cooler weather or later in the day when there is less transpiration from the leaves. To reduce transplant shock, the leaves of trees and shrubs are pruned back. Once seedlings have two to four true leaves and have been hardened off, they can be transplanted into the garden or bigger pots. All need to be watered before and after transplanting. Deciduous plants are moved in the dormant season.

## tree

A tree is a tall, perennial, woody plant. It usually has a single trunk, with branches, twigs and leaves that form a crown.

Deciduous trees lose their leaves in winter and become dormant until spring. Evergreen maintain leaf cover but shed old leaves as young leaves replace them throughout the year.

Trees come in different shapes and sizes, as do their leaves and fruits and seeds. This may help determine what you choose to plant in your garden.

*see also* **clump, fruit trees**

SOME TREE CANOPY SHAPES

canopy of branches and leaves

trunk

acorn

English oak
(*Quercus robur*)

SPREADING

TREE

winged seeds

sugar maple
(*Acer saccharum*)

OVAL

cone

leaves to 3 mm long

winged seed

American elm
(*Ulmus americana*)

VASE-SHAPED

catkin

silver birch
(*Betula pendula*)

PENDULOUS

Lombardy poplar
(*Populus nigra* var. *italica*)

COLUMNAR

Alaska cedar
(*Chamaecyparis nootkatensis*)

CONICAL

**trench composting** *see* dig and drop composting

**trench layering**

All leaves except those at the tip of a low branch or a new plant are removed. It is bent over and pegged horizontally in a trench, then covered with soil.

The horizontal orientation causes new plants to root at nodes along the stems, and grow new plantlets vertically.

Trench layering works best with plants whose buds will grow under soil, like viburnum and some fruit trees

**true leaves** *see* germination

**true to type**

True to type, also called to true to seed. refers to seeds that will grow the same kind of plant as the parent plant.

**tuber**

A tuber is a swollen part of an underground stem or root that stores food.

Underground stem tubers, like potatoes, have nodes with buds and scale leaves on the tuber rather than true leaves. Root tubers like dahlias and sweet potatoes also form buds on their tubers.

Both stem tubers and root tubers can be planted whole as seed tubers or cut into pieces, each with a bud, for propagation.

TUBERS

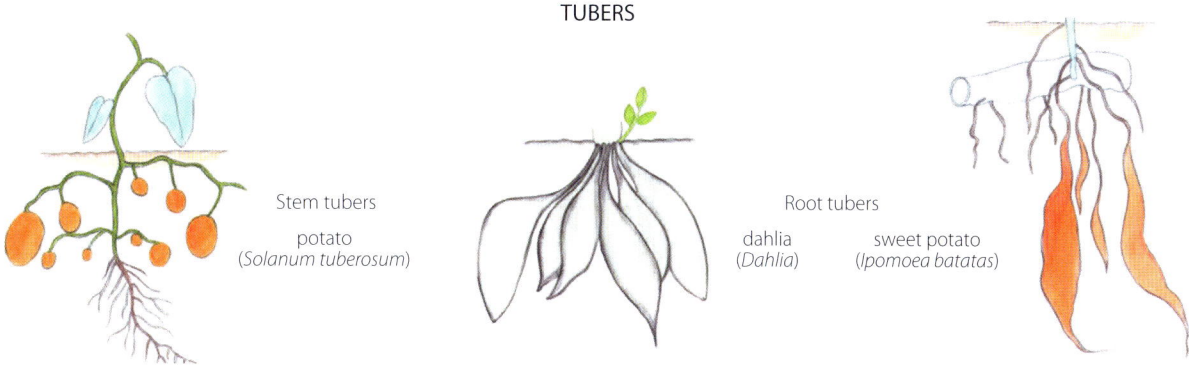

Stem tubers
potato
(*Solanum tuberosum*)

Root tubers
dahlia
(*Dahlia*)

sweet potato
(*Ipomoea batatas*)

**tuberous roots**

A tuberous root is enlarged to function as a storage organ. The enlarged area can be at the end or middle of the root, or it may take up the whole root.

Root tubers of dahlias form buds for propagation where they are attached to the old flowering stem. Sweet potato tubers develop buds on the surface near the stem. Small tubers can be planted as seed tubers. Larger tubers are divided into pieces, each with a bud for propagation.

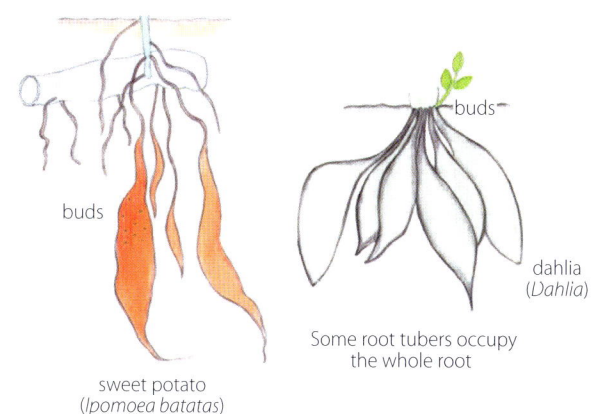

buds

buds

dahlia
(*Dahlia*)

Some root tubers occupy the whole root

sweet potato
(*Ipomoea batatas*)

## tunicate bulbs

Tunicate bulbs, like onion and tulip, have a tight-fitting outer coat (tunic) that encloses the fleshy leaf scales, leaves and buds. All are attached to a short stem, the basal plate, from which the roots, leaves and flowers grow.

The fleshy scales store nutrients and water for the bulb when it starts to grow again. The parent bulb lasts indefinitely and reproduces by seeds and offsets.

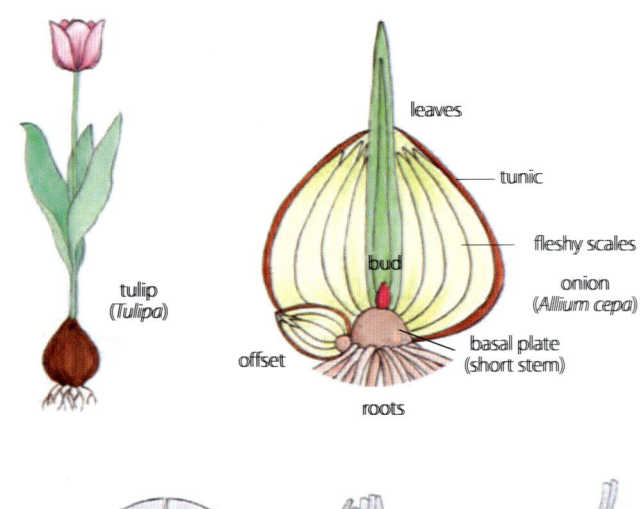

tulip
(*Tulipa*)

leaves

tunic

fleshy scales

bud

onion
(*Allium cepa*)

offset

basal plate
(short stem)

roots

## twin-scaling

Twin-scaling is a time consuming method of propagating bulbs, and it can take several years before the bulblets flower.

The bulb tunic is removed and the roots trimmed so that the bulb can be sterilised by dipping into a dilute bleach. It is then cut vertically, through the scales and basal plate, into sections (chips). The chips are further divided into pairs of scales. The twin scales are placed in a sealed plastic bag with moist vermiculite, or other suitable medium, and stored in a warm dark place until bulblets develop. The bulblets are then planted in pots.

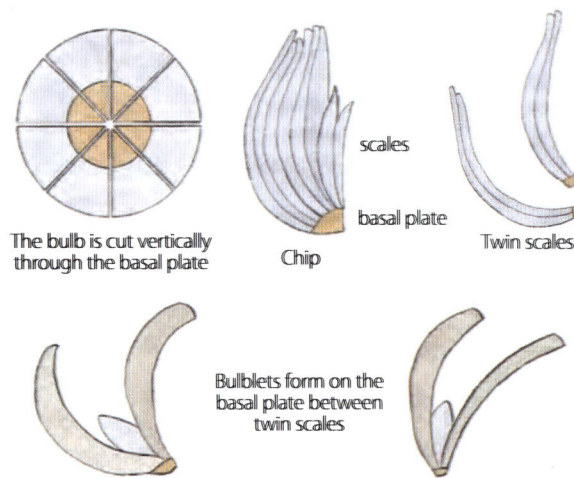

The bulb is cut vertically through the basal plate

scales

basal plate

Chip

Twin scales

Bulblets form on the basal plate between twin scales

## twiners

Twiners, also called bines, have climbing stems that wind around a support. Perennial woody climbers, like bougainvillea, are called lianas.

## undercut pruning

Undercuts are used to remove large branches. The method prevents damage to the tree, in particular to the collar region that plays an important role in healing the wound left behind after pruning.

The first cut, the undercut, is made about 30 cm from the collar and should be a quarter to halfway through the limb.

The second cut is made on the topside about 10 cm away from the undercut. Keep cutting until the branch begins to snap then falls cleanly away, removing the bulk of the branch's weight.

The third cut removes the stub and is made through the branch close to the collar. It is angled so that water will run off the open wound.

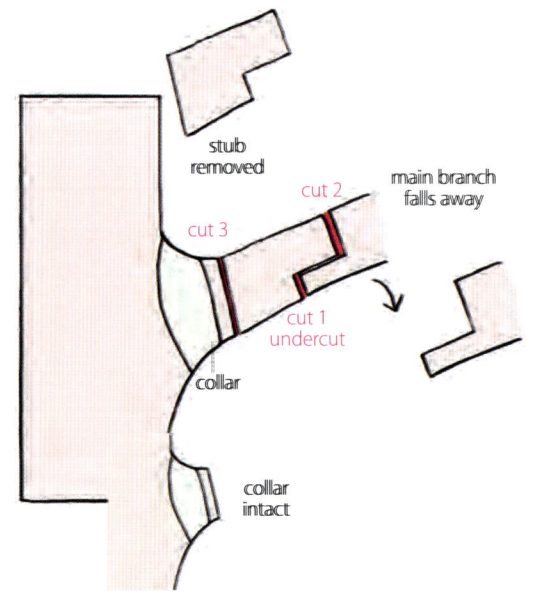

stub removed

main branch falls away

cut 2

cut 3

cut 1
undercut

collar

collar intact

Undercut pruning

## unisexual flowers

Unisexual flowers have pistils and stamens in separate flowers. The female flowers and male flowers may be located on the same plant or on different plants of the same variety.

Pumpkins, cucumbers, papaya, maize and watermelon have male and female flowers on the same plant (monoecious) and can self-pollinate, or cross-pollinate with flowers on another plant.

Hollies and asparagus have male and female flowers on separate plants (dioecious), as do date palms and most apple trees. Since only the female plants can produce fruit, they must have a male plant and a female plant in close proximity.

*see also* **male and female flowers**
*cf.* **bisexual flowers**

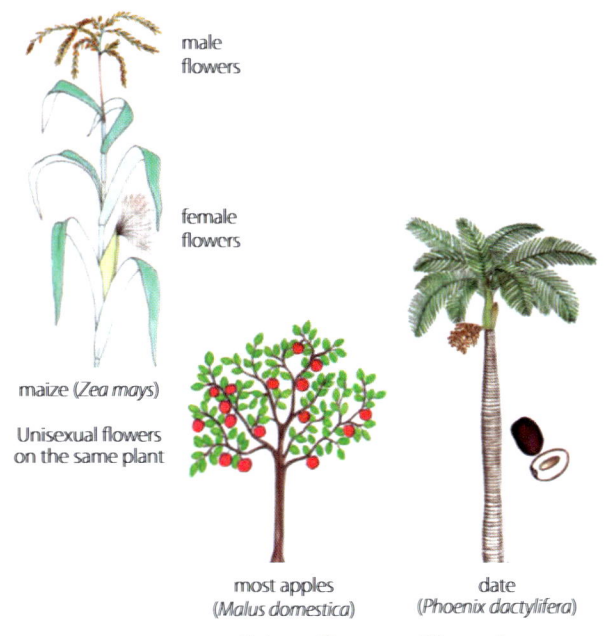

male flowers

female flowers

maize (*Zea mays*)
Unisexual flowers on the same plant

most apples (*Malus domestica*)

date (*Phoenix dactylifera*)

Unisexual flowers on different plants

## urban farming

Urban farms produce food in a city or other heavily populated areas.

Over 90% of Australians live in urban centres with long food supply chains and high supermarket prices for fresh garden produce. Urban farms are becoming popular because they reduce food miles and give access to fresh, affordable food. They are run as a business.

Urban farming can be done in a backyard, greenhouse, on a rooftop, in a shed, or on a vacant lot. Farmers use systems like indoor vertical gardens for mushrooms, hydroponics for growing tomatoes and strawberries, and aeroponics for leafy greens.

*see also* **market gardens, permaculture**

## variegated

A variegated plant has one or more different coloured patches, spots or streaks, commonly on its leaves or petals.

Plants that have variegated leaf varieties include pelargoniums, cyclamens, begonias and hostas.

Variegated pelargonium leaves

**variety** *see* page 144

**vegetative propagation** *see* page 144

**vermicomposting** *see* worm farming

## vermiculite

Vermiculite is a spongy material made by superheating mica. It is used in potting mix to aerate the soil and hold moisture.

## vernalisation

Vernalisation is the process in nature of preparing a plant to flower in spring by exposing it to long periods of cold in winter.

## variety

In botany, variety is a category that ranks below species and subspecies.

Variety and cultivar both refer to variation of a plant within a species. A variety develops naturally, while cultivars are developed artificially by humans. Seeds from a botanical variety tend to grow true to type, while seeds from a cultivar will most likely not grow true to type. The scientific names of varieties and cultivars are written differently.

The term 'variety', in a general sense, is also used to refer to a form of cultivar.

<table>
<tr><td>

**Scientific name:** *Cercis canadensis* var. *texensis*

| Family: | Fabaceae |
|---|---|
| Genus: | *Cercis* |
| Species: | *canadensis* |
| Variety: | var. *texensis* |
| | |
| Common name: | Texas redbud |

</td><td>

**Cultivar name:** *Cercis canadensis* 'Silver Cloud'

| Family: | Fabaceae |
|---|---|
| Genus: | *Cercis* |
| Species: | *canadensis* |
| Cultivar: | 'Silver Cloud' |
| | |
| Common name: | Texas redbud |

</td></tr>
</table>

## vegetative propagation

Propagation is done from the vegetative parts of plants, like leaves, roots and stems, rather than from seeds.

Gardeners propagate plants vegetatively by grafting, layering, division and separation. Micropropagation can be used by garden enthusiasts who are willing to buy special equipment.

Plants that are reproduced vegetatively are genetically identical to the parent.

### VEGETATIVE PROPAGATION

bulb offset    cormels    cutting    tubers    layering

grafting    rhizome    stolon    clump    bulbils    offset

Division    Separation

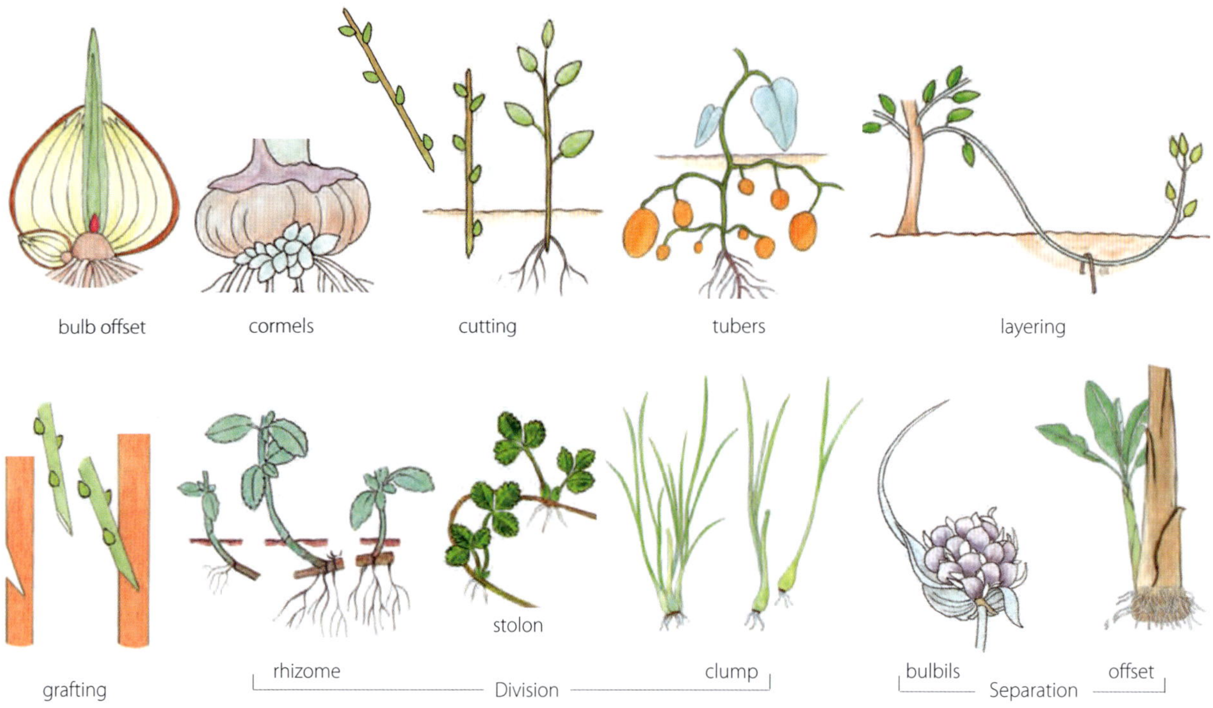

## vine

A vine is a thin-stemmed climber or scrambler that uses tendrils or twining shoots to climb, usually on another plant. The stem itself does not twine.

It may be herbaceous or woody, annual or perennial.

Examples are grape vines (*Vitis vinifera*) that climb by tendrils, and clematis (*Clematis*) that climbs by twining its leaf stems around a support.

clematis                                          grapevine

## water

All water comes from precipitation, such as rain or snow. It is captured naturally on the surface of the Earth in lakes, rivers, creeks and wetlands, and is also stored below the surface as groundwater.

Unsustainable water use that results in overused groundwater, seasonally dry rivers, disappearing lakes and wetlands is a problem in Australia and across the world.

In the 20th century huge dams were built, usually filled from a river system, for use in agriculture, industry and households. This gave a false sense of the amount of water available and how its collection was harming the environment.

The solution for gardens is apparently simple: less and more efficient use of the water available, harvesting of rainwater in tanks and swales, use of treated grey water, water-wise practices like using mulch and planting to suit the climate, and creating a healthy, living soil that will hold water.

## water harvesting *see* water

## water shoot, water sprout

A water shoot or water sprout is a vigorous, slender upright shoot that develops from a dormant bud on the trunk or branch of a tree.

They are usually removed. Water sprouts on fruit trees bear little if any fruit.

Tree with water sprouts on the trunk and a branch

## water tank

Rainwater can be harvested from rooftops on houses and other structures and stored in tanks for watering gardens.

Tanks come in shapes and sizes suited to a variety of spaces.

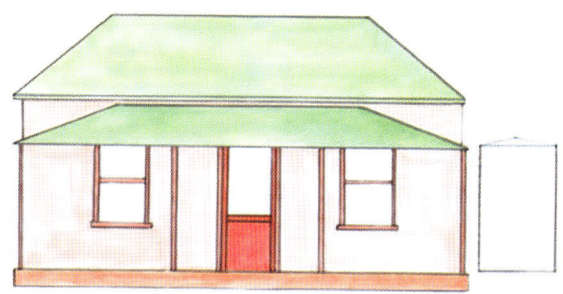

## wedge grafting

Wedge grafting, also called cleft grafting, is done on an existing healthy tree that has been cut back to a stump. Short sticks of 1-year-old wood from a new cultivar (the scions) are inserted (grafted) into a cut made across the stump (the rootstock). The cut is held open with a wedge.

One scion is inserted in a smaller rootstock, or on the edge and aligned with the cambium if the rootstock is wider.

Wedge grafting is used in green-grafting.

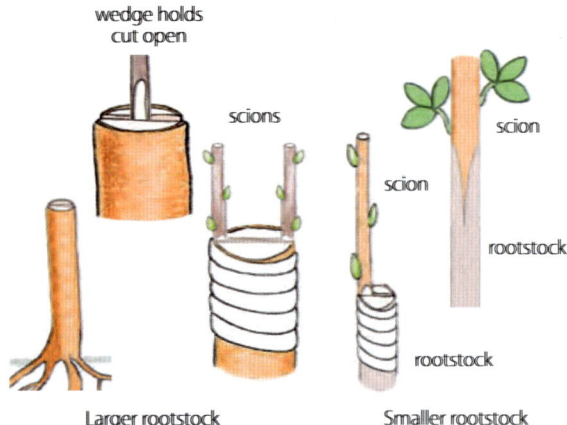

Larger rootstock          Smaller rootstock

## weeds

A weed is an unwanted plant in your garden. Plants that spread easily, like agapanthus (from seeds), freesias and watsonias (from seeds and corms), can escape and become environmental weeds.

'Weeds', like some species of the daisy *Senecio*, are among the first colonisers of disturbed spaces. They are part of a natural succession of plants that re-establish nature on disturbed sites and can eventually be replaced by desirable shrubs and trees. A garden can be considered a disturbed space.

There is a bank of weed seeds buried in the soil. It is best not to disturb them. No-dig gardens have fewer weeds. If flowers and vegetables are planted closely enough, they block light from germinating seeds. Interplanting and cover crops help reduce weed germination too.

Covering bare soil with mulch blocks out the sun and reduces weed germination. Hot composting will kill most weeds. A mulch of good compost, 5–10 cm deep, will feed the soil and prevent weeds sprouting. Some common herbs, like creeping thyme, prostrate oregano and rosemary, can be used as a living mulch.

Solarisation will get rid of weeds from a garden bed before planting.

## whip and tongue grafting

This graft is similar to a splice graft, except that a second 'tongue' cut is made in both the scion and rootstock. This allows the two to fit tightly together.

The cut surfaces expose a large area of vascular cambium for union. The graft is then held together with grafting tape and covered with wax.

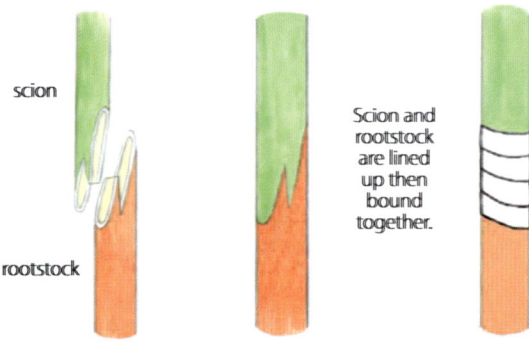

## white fly *see* pests and diseases of gardens, white oil

## white oil

White oil is a home-made pest control spray used to control sucking and chewing insects like aphids, white fly and fruit moths.

It is a mixture of soap or detergent and vegetable oil that kills pests by suffocating them. Test a small area of the plant first. Apply the spray liberally, usually early in the morning, to targeted areas. Spraying can be repeated as often as required. While the spray is harmless, it is a good idea to wash treated fruits and vegetables before eating them.

## wicking beds

A wicking bed has a waterproof liner that prevents water from leaving the bottom of the bed. The water lies below the soil in a layer of scoria or fine gravel. A covering of geotextile prevents the layer of soil above it from mixing with the scoria.

The reservoir of water is filled through a pipe. Water is then drawn upwards through the soil to the roots by capillary action.

The bed can be planted three to five times a year, depending on the crop. Food crops include vegetables, salad greens, herbs, spices and berries. Cut and come again crops are suited to wicking beds.

The system is very water-efficient. There is no need for overhead watering once seedlings are established. There is less water loss through evaporation, especially if the surface is mulched. The bed uses about one-seventh of the water needed for a traditional kitchen garden.

The wicking bed was invented by Australian engineer Colin Austin. These self-contained gardens are now used all over the world and are particularly valuable in arid areas and urban spaces.

**WICKING BED**

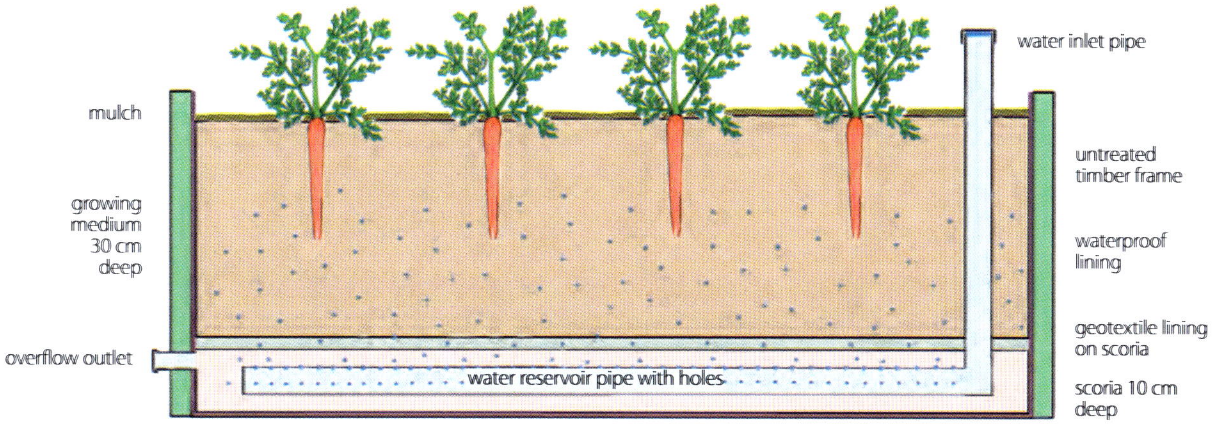

**wilt** *see* pests and diseases of gardens

**wood** *see* xylem

**worm composting** *see* worm farming

**worm farming** *see* page 148

## worm wee

Worm wee is liquid that seeps out from a worm farm. It is sometimes called 'liquid gold' because of its superior qualities as a fertiliser.

It must be diluted to one-part worm wee to ten-parts water so that it looks the colour of weak tea.

## worm farming

Worm farms use surface-dwelling earthworms that efficiently break down organic waste, producing a superior nutrient-rich organic fertiliser as a by-product. The process also provides a sustainable solution for food waste and animal waste management.

Common worms used for processing organic waste are redworms (*Eisenia fetida*) that occur naturally in the organically rich surface layer of the soil where they feed. The common garden worm (*Lumbricus terrestris*) burrows and lives in the soil. They are not suitable for worm farms.

Worms are raised in small or large beds, either above the ground or on the ground. All beds must be aerated, covered and kept moist. A bedding of shredded newspaper, old leaves, aged compost, animal manures, and some sand or soil is set up before the worms are introduced. Worms can consume more than their own weight in organic matter each day from the moment they hatch. They can be fed garden waste, food scraps and aged manures.

Worms, worm castings and worm wee can be harvested and sold as a source of income for community gardens and larger gardens.

*see also* **composting worms, earthworms**

### TYPES OF WORM FARMS

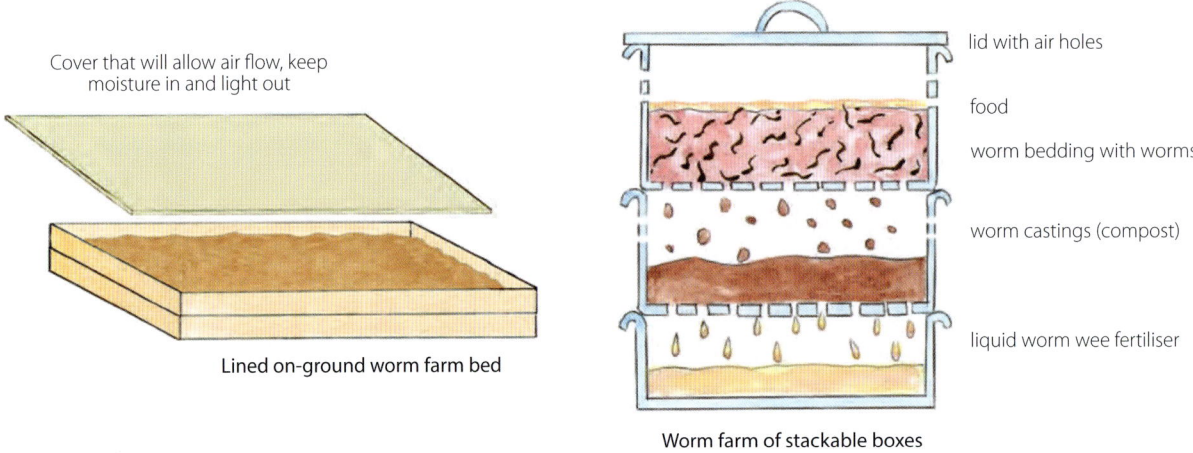

Cover that will allow air flow, keep moisture in and light out

Lined on-ground worm farm bed

lid with air holes

food

worm bedding with worms

worm castings (compost)

liquid worm wee fertiliser

Worm farm of stackable boxes

## wounding

In propagation, wounding is the removal of a thin slice of bark to expose the cambium layer that will develop roots. The wound is dusted with rooting hormone.

A cutting has a short strip of bark removed from the base. In layering, a shallow incision is made on the underside of a node on the stem.

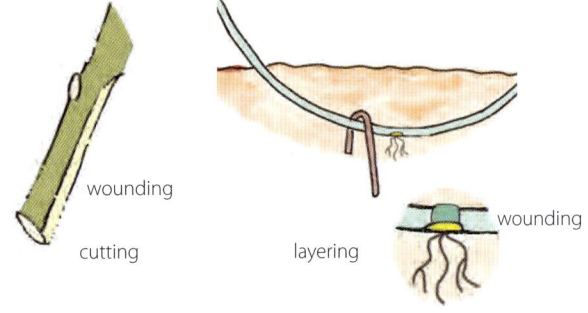

wounding

cutting

layering

wounding

## xylem

Xylem is the wood in woody plants.

Sapwood is living xylem. It is a cylinder of living cells that runs the length of the plant and transports nutrients and water from the roots to all other parts of the plant.

Dead xylem is called heartwood. It was once sapwood, and when it is no longer functioning it becomes the strong central pillar of a tree.

outer bark
inner bark (phloem)
cambium
sapwood (living xylem)
heartwood (dead xylem)

**zinc** *see* plant nutrients

---

# A selection of references

## Garden history

Carroll-Spillecke M (1992) The gardens of Greece from Homeric to Roman times. *The Journal of Garden History,12(2)*, 84–101. <https://doi.org/10.1080/01445170.1992.10410564>

Engles J (2020) Ancient gardens of North America <https://www.permaculturenews.org/2020/09/29/ancient-gardens-of-the-north-america/> Permaculture News, Permaculture Research Institute

FUNCI (Islamic Cultural Foundation) (2012, posted online 2017) Gardens of Al Andalus <https://funci.org/gardens-of-al-andalus/?lang=en>

Garden Visit (accessed June 2024). West Asian gardens <https://www.gardenvisit.com/history_theory/garden_landscape_design_articles/west_asia>

Gothein ML (1928) *History of Garden Art* (2 volumes). JM Dent & Sons, London

Government of Mexico City (2017) Chinampa Agricultural System of Mexico City, Mexico <https://www.biopasos.com/biblioteca/Chinampa-agricultural-system-mexicocity.pdf>

Hill AW (1915) The history and functions of botanic gardens. *Annals of the Missouri Botanical Garden*, **2(1/2)**, 185–240. <https://doi.org/10.2307/2990033>

Med-O-Med (accessed June 2024). Cultural landscapes of the Mediterranean and Middle East <https://medomed.org/>

Muslim Heritage (accessed June 2024). Agriculture <https://muslimheritage.com/category/environment/agriculture/>

Nematdjonovna SS (2019) 'CHAR-BAGH' Gardens through the ages. *International Journal of Advanced Research in Science, Engineering and Technology* **6 (10),** 11372–11380 <http://www.ijarset.com/upload/2019/october/69-sitarasdyova-44.pdf>

Sellers, VB (2003). Gardens of Western Europe, 1600–1800. In *Heilbrunn Timeline of Art History. New York: The Metropolitan Museum of Art, 2000–*. <http://www.metmuseum.org/toah/hd/gard_1/hd_gard_1.htm>

Sheldon N (2018) Ancient Roman gardens. History and Archaeology Online <https://historyandarchaeologyonline.com/ancient-roman-gardens/>

The British Museum (2018) Paradise on earth: the gardens of Ashurbanipal <https://www.britishmuseum.org/blog/paradise-earth-gardens-ashurbanipal>

## Gardening practices

AusVeg (accessed June 2024) Overview: Pests, diseases and disorders <https://ausveg.com.au/biosecurity-agrichemical/crop-protection/overview-pests-diseases-disorders/>

Barth B (2018) How does aeroponics work? <https://modernfarmer.com/2018/07/how-does-aeroponics-work/>

Charles Dowding (2004) Advice for starting no dig [gardens] <https://charlesdowding.co.uk/start-no-dig/>

McCallum Made (accessed June 2024) Ecological gardening for growing food <https://www.thechickentractor.com.au/ecological-gardening-for-growing-food/>

Welsh DF, Janne E (2008) Follow proper pruning techniques. Texas AgriLife Extension Service <https://aggie-horticulture.tamu.edu/earthkind/landscape/proper-pruning-techniques/>

Homestead.Org (2019) The basics of biodynamic gardening <https://www.homestead.org/gardening/the-basics-of-biodynamic-gardening/>

NRDC (Natural Resources Defense Council) & Hu S (2020) Composting 101 <https://www.nrdc.org/stories/composting-101>

Permaculture Visions (2018) Difference between organic gardening and permaculture. <https://permaculturevisions.com/difference-between-organic-gardening-and-permaculture/>

Plants Bank (accessed June 2024). Understanding hydroponics, How to grow plants in water <https://plantsbank.com/understanding-hydroponics-how-to-grow-plants-in-water/>

Vinje E (2023) Plant Propagation 101. Planet Natural Research Center <https://www.planetnatural.com/plant-propagation/>

Wallace EJ (2019) The Moroccan food forest that inspired an agricultural revolution <https://www.atlasobscura.com/articles/what-is-permaculture-food-forests>

## General

BioCycle: The Organics Recycling Authority

Dayleys Fruit Tree Nursery <https://www.daleysfruit.com.au/>

Deep Green Permaculture<https://deepgreenpermaculture.com/>

Epic Gardening <https://www.epicgardening.com/>

Gardening Know How (US-based) <https://www.gardeningknowhow.com>

Geoff Lawton, Permaculture Online <https://www.discoverpermaculture.com/> [Note: you are requested to complete a free registration to access this content]

Hemenway T (2009) *Gaia's Garden: A Guide to Home-Scale Permaculture*. 2nd Edition. Chelsea Green Publishing Co, White River Junction, VT

North Carolina State University (2020) *North Carolina Extension Gardener Handbook*. 2nd Edn. (Eds. Moore KA & Bradley LK). <https://content.ces.ncsu.edu/extension-gardener-handbook>

Oregon State University, OSU Extension Service <https://extension.oregonstate.edu/>

Permaculture Research Institute <https://www.permaculturenews.org/>

Planet Natural <https://www.planetnatural.com/>

Rodale Institute <https://rodaleinstitute.org/>

SGA (Sustainable Gardening Australia) <https://www.sgaonline.org.au/>

Simplify Gardening <https://simplifygardening.com/>

Soil and Health Library <https://soilandhealth.org/>

Tenth Acre Farm: Permaculture for the Suburbs <https://www.tenthacrefarm.com/>

The Spruce (US-based) <https://www.thespruce.com>

VicFlora (2024) *Flora of Victoria, Royal Botanic Gardens Victoria* <https://vicflora.rbg.vic.gov.au>

Help Me Compost <https://helpmecompost.com/>